HAMID AWONG FISHERIES MODEL (HAFM):

A CASE STUDY STOCK ASSESMENTS OF DEMERSAL FISHES OF PRIACANTHUS TAYENUS (RICHARDSON 1846) IN DARVEL BAY, SABAH, MALAYSIA

BY

HAMID AWONG

© MMVIII. All Rights Reserved.

This book is printed on acid-free paper.

Copyright © 2008 Hamid Awong. All rights reserved.

Published by Aberdeen University Press Services.

No part of this book shall be reproduced, stored in a retrieval system, or transmitted by any means, electronic, mechanical, photocopying, recording, or otherwise, without written permission from the publisher. No patent liability is assumed with respect to the use of the information contained herein. Although every precaution has been taken in the preparation of this book, the publisher assumes no responsibility for errors or omissions.

Every effort has been made to make this book as complete and as accurate as possible, but no warranty or fitness is implied. This information provided is on an "as is" basis. The publisher shall have neither liability nor responsibility to any person or entity with respect to any misguidance or misunderstanding from the information contained in the book.

Printed in the United States of America

ISBN: 978-0-6152-1321-7

This book is printed on 8" x 11", perfect binding, 60# cream interior paper, black and white interior ink, 100# white exterior paper, full-color exterior ink. Prices are subject to change.

Cover Title Designed by Aberdeen University Press Services.

Hamid Awong Fisheries Model (Hafm): A Case Study Stock Assesments of Demersal Fishes of Priacanthus Tayenus (Richardson 1846) In Darvel Bay, Sabah, Malaysia
First Edition

Hamid Awong

This book is dedicated to

My beloved Wife, Ir.Dayang Salmah Datuk Hj. Sidek, University Malaysia Terangganu
Daughters; Mahar Diana, University Sheffield, United Kingdom,
Komala Intan, University of Staffordshire, United Kingdom,
Salha Pentawinata, International University College TwinTech, Kuala Lumpur,
Hexanti Ayu, Lahad Datu, Sabah,
Azizah, Lahad Datu, Sabah,
Sons; Mohshie, University Malaya, Kuala Lumpur
Basyid, University Malaysia Sabah

- Hamid Awong

ABSTRACT

THE FISHERIES MODEL (HAFM): A CASE STUDY STOCK ASSESMENTS OF DEMERSAL FISHES OF Priacanthus Tayenus (Richardson 1846) IN DARVEL BAY, SABAH, MALAYSIA

The species of *Priacanthus tayenus* (Richardson, 1846), Threadfin big eye or known as Lolong Bara in Malaysian Peninsula and Beliak Mata in Sabah are mostly found in Darvel Bay, Sabah. The fishery stock assessment in Darvel bay was specifically studied from various aspects of fishery. This study was divided in four stages. The firstly stage was focused on stock assessments including the ratio of catch such as composition, marketable, underutilized and the fish species composition. Secondly, the biomass dynamics and exploitation level of Threadfin big eye (*Priacanthus tayenus*) by trawler in Sabah. The thirdly was the population dynamics of *Priacanthus tayenus* and last the Status Model of Threadfin big eye *Priacanthus tayenus* within a period 50 years started from 2003 to 2052.

The survey area was conducted in Darvel Bay, and the trawlable fishing area was 834.55 kilometers square. The study on demersal resources by using Swept Area Method had shown that the potential resources were 1,034,979.15 kilograms. The estimated density in this area was. 1,240.16 kilograms or 1.2 metric tones per square kilometer. There were 128 hauls taken during the trial with a total of 20,032.9 kilograms of fish. Caught species identification was done and there were 71 species of fish caught in the trial. The composition ratio of the catch between marketable and underutilized species was 12,426.9 kilograms or 62.03% kilograms trash fish while marketable species was 7,606 kilograms or 37.96%, where the average catches were 39.12 kilograms per hour.

A status "model HAFM" (Hamid Awong Fisheries Model) was established to determine the population stock, yield (biomass) and the value of the stock level started from 2003 until 2052. This model was used to show the relationship of biology parameter from previous chapter between recruitment, growth and mortality total either mortality, fishing mortality or natural mortality of *Priacanthus tayenus* in Darvel Bay. Assumed the initials number of stock in 2003 was 100 heads. The number of stock fall into the lowest level, the recruitment failed to recover with current effort. In 2052 or 50 years later, the expected population will be 1.7×10^{-8} while the yield (biomass) will be 4.0×10^{-7} gram and the value for the stock RM 2.2×10^{-6}.

TABLE OF CONTENTS

S/N	TITLES	PAGE
1.	Introduction	1
2.	Literature Review	15
3.	Description of the Study Area	61
4.	Materials and Methods	71
5.	Survey on the Demersal Marine Fishes by Trawler	93
6.	Hamid Awong Fisheries Model (HAFM)	107
7.	Conclusion and Recommendations	161
8.	References and Bibliography	169
9.	Appendex	179
10.	Lists of Abbreviations	201
11.	About Author	203
12.	Index	205

INTRODUCTION

1.1 Background of the Study

An increase in world population is proportional to food supplement. Fishes are among food for most of the people. They are increasingly demanded which have to be met by fully utilization of all fish species in a sustainable manner. At present the fish stocks in most coastal states especially in the developing countries are being exploited beyond Maximum Sustainable Yield (MSY) or over exploited and over capitalized where revenues cannot cover the cost. The result of these activities caused lost of profits to the fisherman. Apart from that, the resources are not fully utilized, lack of skill in resources management, processing and handling of fish identification which caused resources depletion of the stocks.

Today, demand of fish as source of food for human consumption is high and there are important link with the food chain for fishes. Nature maintains a balance among the fish population as a whole as to catch the fishes continuously. Governments of various countries are concerned with control of the industry to avoid reduction of fish population.

Fisheries Resources

Fisheries is a renewable resource. It has reproductive capability as source of food for human and animal consumption. There are about 21000 species of fish or half of the animal backbone in the world (King, 1995). These animals are distributed from higher land into deep seas and from tropics to Antarctica. The catch trend increases by quantity every year because of increasing effort using fishing gear. It had been claimed that there was an increased in the total world catch of fish from an estimation of 2 million tones a year in the nine teenth century to approximately 75 million tones a year in early 1980. This was made possible by using mechanical power and electronic aids to harvesting, processing and distribution (Eddie, 1980). Higher technology in fishing also led to increase fishing efficiency and fishing power.

The annual growth of fish has amounted to about 1000 million tons (Gaskell, 1970) where the world catch is approximately 30 million tons per year. In this case, there are still plenty of rooms to expand the world fishing industries and need to understand the effect of fishing activities for sustainable purpose.

The fish consumption in Malaysia is 60 kilogram per capita (Berita Harian, 21 August 2003). It had become a primer food for most of the people in the developing or under develop countries. Fish is can be taken for breakfast, lunch and dinner or as snack food. The fish processing gradually will cause stock depletion. As consequences, these can create food problems to the most people in the coastal state. (Sabah Annually Report, 2000)

Sabah as a coastal state has about 500 species of fish available. It is important to assess the potential biomass and identify the major species that have commercial value or commercial species for conservation purposes. The catches by most fishing gear are very low and mostly on non-targeted fish.

Fisheries Management

The fisheries management is a tool to control resources. Fisheries sectors are important to the economic development in every developed or developing country. The society living in the coastal state obtain food as well as income from the fisheries. Better fisheries management is needed to sustain the fisheries stock.

This phenomenon of depletion of fisheries resources is needed in order to solve the problems becoming big and serious. The manager should protect the stock from being over exploited by reducing the catch. The situation has led the fishing industries economically marginal because the stock is being depleted. Environmental damage occurred because of unhealthy fishing practice such as using dynamite and cyanide poisoning. In future, global conservation standard and guideline focusing on long term fisheries management plans for all fisheries is needed.

On the other hand, increase in levels of technology are not always accompanied by correspond rises in productivity. The most dramatic Illustration was a case where improved technology had an impact to the catch rates of *Peruvian anchoveta* (notably power block purse seining and the use of synthetic nets).

Management of marine resources and identification of commercial species and other species are much needed. Besides, the study of surplus yield or biomass dynamic is useful as a guideline for the economist to utilize the resources with sustainable manner. In order to identify the exploitations level, it is important to evaluate the biomass status and the number of fishing fleet involves. These collected data could be used for sustainable purpose in further management. Trawler is identified as an active gear that can catch all type of fish especially demersal fish.

The biological information data could be used to guess or forecast the stock and yield from the stock resources in that area for certain time. This information is useful for fishery manager to manage the resources in future.

Maximum Sustainable Yield

Fisheries sector are important and popular for economic development in every developed or developing country. Food can be processed from fisheries resource. Most of the society in the coastal state depend their income from fisheries. In the other aspect the fish stock are being exploited beyond the Maximum Sustainable yield (MSY) or over exploited and over capitalized where revenue can not cover the cost and as a result, the fishermen lost their profit. In order to avoid depletion of fisheries stock, the manager should include a program for recruitment in that area. This is because the stock will not be recovered by the recruitment of stock.

This phenomenon needed to solve the problem from getting bigger and serious. It is a task to the manager to protect the stock being over exploited by reducing by catch problem. Other alternative is to reduce too many people depend on fisheries for their live hoods. This situation would lead the fishing industries economically marginal because the stock is being depleted. The environmental damage because of fishing technologist and unhealthy fishing practice such as using dynamite and cyanide poisoning. In future standard global conservation and guideline focusing on long term fisheries management plans is needed for all fisheries.

Fisheries Development in Sabah

Sabah is situated in North Borneo. Sabah has a coastline area of 1,440 kilometer or 900 miles. The island is washed by the South China Sea on the West, the Sulu Sea and Celebes Sea on the East. The land area is approximately 74,500 Kilometer square or 29,399 miles square. Lying on north of the equator with tropical climate it has summer all of the year with daily temperature ranging from 23 to 33^0 C.

Sabah has a population about 2 million comprises of 47 different races and speaking with their own dialect. 30 percent of the population is full time or part time fishermen. It is about 40 000 fishermen in Sabah. It plays as a major role to the state economic resources and also as a major source of protein. Sabah has pure white beach sand where the island of Sipadan located off to the east coast. It has one of the best diving spots in the world rising 600 meters on a limestone pinnacle out near the surface surrounded with abundance of marine resources

The fishing ground is divided by zone. There are four marine zones such as Tawau, Sandakan, Kudat and West Coast. Fishing ground in Sabah can be divided into three main zones namely West coast facing to the South China Sea, Kudat Region facing to the

South China Sea and Sulu Sea, East Coast facing to the Sulu Sea and Sulawesi Sea. The trawler main area is West coast Kudat and some in the east coast.

The regions are divided into 16 marine districts to oversee all aspects of fisheries management such as administrative, planning development social economic project, data collecting related to fisheries, landing information, enforcement and utilization of fisheries resources.

Unfortunately, less fisheries information data is available throughout the State. This is due to lack of research in the past. In order to avoid the resources from being depleted and over exploited, fisheries information data is required for conservation purpose in the future.

Unselective fishing gear will cause endanger to the resources. Better management had been applied to control the resources. Enforcement has been implemented throughout the state largely to monitor the fishing activities. Sabah Fisheries Department is responsible in the management of fish stocks in the state.

Fishing gear can be divided into commercial and artisenal. Commercial fishing gear includes purse seining, trawler and active gear while the arisenal fishing gear is a traditional fisher with limited cost and limited catch such as selambau, kelong, pole and line gill net trap. Bag net is a fishing gear with the main target is small pelagic and mostly concentrated in East Coast.

Trawler is a commercial fishing gear that catches multispecies of demersal fish. The trawler range from non-powerboat to below 70 gross tones is used mostly as fishing gear. Trawler is used as a source of fishmeal plant. At the moment there are six fishmeal plants in the state (Sabah Fisheries Annual Report, 2000). Trawler was first introduced in the state during the sixties. The net mesh size with code end diamond size was 38mm, which was legally accepted by the fisheries authority at that time. The net was operated using double otter beam boat with main target the species living in the bottom of the sea. The specified area should be mud for trawler able, most of the operation area is within 12 nautical miles along coastal line. The trip to the sea will take 3 to 4 days. The east coast of Sabah is known as trawler able most of the year. The trash fish or by catch fish using trawler is considered as underutilized because of their under market size. It is in the category of commercial species, but has uneconomic prize at the rulings market prize level.

The fisheries stock in Sabah can be divided into two main categories, pelagic and demersal fish. Fishing area is in the coastal zone with in 19.2 kilometers (12 mile nautical) and beyond19.2 kilometers (12 mile nautical) 19.2 kilometers (12 mile nautical). The fish species being caught include *Priachantidae spp. Nemipterus spp.* or the other are *Decapterus spp. Mullidae spp. Lutjanidae spp.* and other commercial fishes.

The cost to build a vessel is relative high that only can be effort by established fisher. The crew consists of 3-4 people in one trip. The fishing ground for trawler in the east coast is limited due to the boundary with other countries as Philippine and Indonesia. Jetty was built close to the market at most of the town in the state is useful for the catch landing. The price depends on the market that is subject to demand and supply in that area.

The high demand of fish for food is high has a correlation to rapid population growth and increasing of immigrants in the State. Costumers are able to buy fresh fish as a cause of better economic growth in the country. Unprocessed and fresh fish could be found in the fish market in every town or even village especially near to the coastal area. The prize in the village is cheaper than in town, which involve the cost for maintenance. Small amount has undergone processing as dry or salted fish while the undersize fish is processed in the fishmeal as pallet for animal feed.

Due to under market size and genuine underutilized resource, the prize at the ruling market is considered low and uneconomic and often most of the small fish is discarded to the sea. Besides, the boat limited room to storage the product. The production cost is high when compare to the ruling prize. None of these fish go through secondary processing to produce new product due to lack of skill. In daily practice, most of the villager processes the product by traditional method such as drying under the sun. This method of processing is low of hygiene and sometimes the product is spoiled by insect and will produce low quality product. As consequences less profit is obtained because of the lower price.

In every town or small town in Sabah there is a daily place or market for local fish products such as salted or dried fish. This fish can also be found in stalls or sent to plantation camp for plantations labors for consumption. The cost to produce salt is simple because it's only involves skill. Sufficient sunlight in the whole year is needed for processing dried fish. Fishermen or their family considers the activity as second activity.

The fresh unprocessed fish has potential for market. The customers always prefer fresh seafood. Usually the fishermen find middlemen to dispose the collected fishes. They are obliged to do so since they are not provided a permit to sell directly to the consumer. These middlemen are gaining more profit than fishermen them selves. In this case, marketing structure should be imposed for the welfares of the fishermen.

The good quality fish were directly sent to restaurants or exported to other districts or other states and countries. This activity is deal by the big traders who have big capital, depend on the size of business. These commodities go through port or airport at major towns in the state. The fishes were in the form of frozen or live fishes. The destinations are peninsular Malaysia, Singapore and Hong Kong.

Darvel Bay

The fisheries in Sabah play a major role on the state economic resources. Totally there are 40 000 fishermen including part timer fisher in Sabah and 1400 fisher in Darvel Bay. Fish is considered as domestic consumption and for export. The resource has become the state and fishermen income. Apart for economic reason the fish is also as a major source of protein for the people in Sabah. Due to the growth of population, higher income and incoming immigrants, the demand of fish as food is high.

Priacanthus tayenus **(Richardson, 1846)**

In Sabah, the *Priacanthus tayenus* (Richardson, 1846) or threadfin big eye is considered as among the fish species, which have a commercial value because of its white meat. It is also considered as superior meat, which is very popular to the local community as food. This species appear to be high abundant and is important as a source of protein for the people. It can be eaten fresh or processed such as salted, brining, dried fish or fermented through secondary processing.

Priacanthus tayenus are scaly fish, a dominant landing species in Sabah. Because of its important economic reason, It is needed to know the present information and status regarding the species. The fisheries are being managed under output and input control. Nevertheless, the management plan should include the catch and recruitment to asses the stock and forecast planning in the future.

1.2 Statement of the Problem

Fisheries Resources in Sabah

The fish commonly caught by trawler are mostly underutilized species because of their small sizes and immature. Thus, most of the catch using a trawler has no market value. The underutilized of small size and immature species are found highly abundant in Sabah water. Because of their smaller sizes, all are discarded to the sea or sent to the fishmeal plant to be processed as animal food. Human is consuming the catch minimal and as a result the resources will be wasted. In this case, catch-using trawler is not applicable and endanger the species.

Decrease of fish stock means a corresponding decrease in the income of bigger group. This will create social problem to the artisenal fisher who are majority in the state. Trawlers are also harmful to the other fishing gear especially traditional fishing gear such as net, pole and line traps or even small boat, which can be swept by the trawler net. For sustainable purpose it is important to verify the number of trawler being operated in Sabah.

Worldwide has shown that when there is an open access to marine resources, there are not so many individual harvesting the resources to conserve fish stock. Competitions amongst fishermen, result to depletion of the stock resources, and the end minimize returns to fishermen or the coastal community.

Small fish of *Priacanthus tayenus* are locally known in this district as Kandaman and in Sabah as Beliak mata while in Semananjung Malaysia known as Lolong bara. It is found in high abundance in the state and is a very important economic source for the fishermen who are depending on this species as their income. Lack of these species will affect the life of fishermen as well as the coastal community for its serves as sources of protein. Thus, state foreign exchange rate also decreased.

1.3   Significance of the study

Due to the problem faced by the fisheries sector in Darvel Bay there is a need to solve the problem to avoid over exploitation of capitalization investment in fisheries sector and depletion of the stock resources in the future.

The resources management stands primarily stands from the common property of nature living marine resources. Both the government and also the scientists are responsible to design a scientific method to assess the stock of marine fisheries in order to manage and protect marine resources in Darvel Bay, and to ensure the utilization of the resources for the best of interests the community.

The status of demersal fish stock in Darvel Bay is a need to ensure that marine resources can be used in an ecologically sustainable manner as efficient as possible. Besides yielding a return to the community, it ensures the benefits in the maximum use of living marine resources within the community.

The high abundance of exploitation levels should be noted. Without proper controlled can lead to depletion of the species and the state will face poor resources in the future. The exploitation of fishing gear especially, the active and commercial fishing gears and the biomass dynamics of *Priacanthus tayenus* or thereadifn big eye should be minimized. Identification of most abundant species of *Priacanthus tayenus* or thereadifn big eye as the main source of protein in the State is very important. If handling and processing of the fish were done properly, the product has a potential for export to earn foreign exchange and as economic catalyst in the state.

To study the biologically parameter means to introduce proper management in future. The species is considered dominant in this area and most of the food and protein sources depend on the species. This species are known, to be living on the sea floor then it's vulnerable to fishing gear such as trawler, which swept the bottom of the sea.

Without proper management and fishing having no size limit could decrease the species population and at the end will cause depletion. Once depletion occurs fishermen will find difficulty to catch the fish. If this is happened, the fish prizes become higher while the catch is small and not profitable. The cost for maintenance fishing gear and vessel will increase which could affect to the State economic.

The population dynamics of *Priacanthus tayenus* or thereadifn big eye in Darvel Bay, help to support collecting data to create the biomass model for *Priacanthus tayenus* or threadfin big eye in that area. The biological data such as annual mortality is important that effect sort and long-term population of this species. When the mortality rate is higher in one cohort means only small amount could survive to enter fishery recruitment and reproduction for the next cohort or generation.

The maximum length of this species can be as a guide for the fisheries manager to manage the time of the first capture. The mature fish will produce profit to the fishermen compare to big quantity of under market size catch. The under market size product has low value or even discarded to the sea and will cause a waste to the resources. The fish biomass will be decreasing where the recruitment could not be done because lack of parental to produce juvenile.

The combination of three factors such as length, weight and age can predict the future population base on annual mortality such as fishing and natural mortality by the fisheries manager. Fishing mortality is caused by catch fishing gear and natural mortality is due to the environmental condition factor or cannibal as a predator by other fish species or larger fish.

The fisherman manager should know how to predict the age of fish according to size and weight from the first capture. It is important in order to gain profit and to avoid collapse in fishery where fishermen will suffer and lost since their investment are not profitable.

Proper control of fishery input will produce good result. When fish is found with high abundant, high selectivity fishing gear should be used to avoid depletion to the resources. There are few fishing gear considered dangerous to the environments because of low selectivity that will destroy the habitat of fish.

The dynamic gear is dangerous compare to the static fishing gear because this gear could move around and swept the school of the fish whatever small or big size. The trawler is considered to be dangerous to the fishery. Some countries banned this fishing gear after study and take into consideration the catch concerning to the resources, with contradiction to sustainability oriented with profit come later.

The most important to avoid depletion and over fishing to ensure the public realise the responsibly to marine especially fish resources. Besides, accept the ideology that resources belong to the public not belong to the group of people such as fishermen. The resources belong to the every community and everybody is responsible its awareness. Everyone should concern to the resources by avoiding member of the public as thief or fishing without proper gear or without permit form authority or local government agency. The enforcement also takes seriously and concern with the resources. The law had been made since colonial 40 year ago. However because of lack of implement, most of the coral in the state had been destroyed as reported that 1 million coral that have potential for breeding and fish habitat had been destroyed.

Some fisher believes with the assumption of higher investments in fisheries industry would gain more or bigger profit. However, this assumption is not correct. It is important to understand that the fisheries can only increase their biomass if the parental can produce juvenile for recruitment new cohort when the fishery resource mature. When investments expand without considering the effect, the resources will collapse and could not recover and finally depletion occur in the fishery.

Most people do not concern about using dynamite for fishing whether they are fishermen or not, the motive of course to obtain instant profit. At the same time it would destroy the habitat and finally no new generation is available for recruitment. This habit can be considered as a thief because of using illegal fishing. The public should aware on the activity and should understand the fisheries resource is a public property not belongs to certain person or group. In order to catch the fish one should apply a permit and pay the fees as a rent.

Investment without control of input or fishing vessel and using unselective net may cause depletion to the resources. Harvesting without limit may reduce the recruitment in the fishery. In the same case when the control was done in the first capture. This can be done by limiting the mesh size of the net that allow only certain level such as mature in market size or the fishermen who is fishing under limit should sent the fish back to the sea to give chance for further recruitment.

Its normal at any where there are group of community take advantages to look after for their own interest or selfish. On the other hand, the authority concerned should not entertain this type of people who ignore the importance of resources for next generation.

Some of the fishery policies worldwide would make the people confuse when in the policies includes the objective for sustainable purpose, source of protein for the people and to get the foreign exchange or to maximize profit. This can be happened because some of the fisheries manager misunderstands about the fisheries resources. A good example when the manager get confuse about the sustainable concept and then make the people confuse. This is happened because the manager fails to confide public the importance of fisheries resource.

Due to the problem faced by fisheries sector in Sabah. It is needed to carry out research, to solve the problem and also manage the resource in the future. Apart from that, this research would also provide biology information data of *priacanthus tayenus* for further management plan.

1.4 Objectives of the Study

1.4.1 To study stock assessment of demersal fish by species survey and population density using swept area method in Darvel Bay.

1.4.2 To establish Hamid Awong Fisheries Model (HAFM) for assess the the fisheries stock.

LITERATURE REVIEW

2.1 Living Marine Resources

Fish and other marine resources are basically source of food for human especially in the coastal state. The coastal community globally needs fish as a source of protein. Developed countries had applied new technology in the industry in order to increase the utilization of fish and other marine resources for human consumption. Sophisticated processing technology was invented to produce various high nutritional seafood as well as safe to be consumed.

In new millennium, as the growth population increase, the demand for food will also be increased. In order to meet the demand of food for the population there must be a production of seafood from marine resources. The sea and ocean has been exploited to harvest these resources. This phenomenon has attracted substantial international community's interest in the marine resources as a food source. The government and other agencies involved in fisheries sector had set up their research development center to analyze and create new product and had prospect to obtain maximum utilization of this marine resources. In the past, these resources become underutilized due to lack of experts to produce and processed this commodity for human consumption. These resources were largely used as a source of food. At present especially in underdeveloped coastal countries the ratio between utilizations and not utilization were considered high because most of the catch unable to meet demand of market size and the catch was non targeted species.

Oceans and seawater, which cover two-thirds of the Earth's surface area, contain vast resources of food for human consumptions for energy and protein. Energy and minerals are also invaluable for human benefit in many indirect ways. World fisheries expanded steadily from World War 11 until now in new millennium age, particularly the fisheries in Asia where the population increased and other eastern-block countries. An annual harvest seems to become more stationary which are approximately 100 million tones, (King.

1996). The catch from the sea, ocean, lakes, ponds and rivers are principally coming from Asian countries. Ocean is the main contribution on the major fish harvest.

Marine resources show a district pattern, favoring the eastern edge of the ocean, the high latitudes through seasonal mixing and recycling of nutrients, and the proximity of large rivers. The estimation of the potential yield vary widely, from a technically optimistic high of 2000 000 000 tones yearly to a pragmatically pessimistic low near the current level (Parker, S. (ed.) 1980).

According to Meadows *et. al* (1988), the present rate of resource exploitation and exponential population growth will result in a collapse of the world's economic system. Regarding the food production per capita, the over exploitation or depletion of many of the world's most valuable fisheries resource is wide spread and increase despite crude attempts for management, due to the increasing of world population. Therefore, the overall potential for food production from the sea has been revived (Bell. 1978).

Living marine resources life in the sea ultimately depends upon the plankton, fishes consume plankton as being beneficial for them. A study in the distribution of plankton is useful to determine the distribution of fish and there are positive correlations between them.

2.2   Fish

Fish exhibit great diversity in ecology; live in waters with average temperature of   less than zero to over 30^0 C. There are about 25000 valid species and has it has been predicted there are 28 500 extant species in total (Jennings, *et al.* 2001) These species have been classified into 482 families and most of them live permanently in the sea.

Fish have many different shapes but most of them have torpedo-shaped and others are round, flat, and other are angular. The smallest at sizes less than 15 mm is dwarf pygmy goby *(Eviota)*, the species from Philippines to the biggest one such as whale shark *(Rhhincodon)* with length 21 meter and weight 25 tons or more. Fish are cool-blooded animals, with backbones, gills and fins and primarily dependent on water as a medium in which to live.

Fishes come from vertebrates and the most numerous in their groups with an estimation of 20,000 recent species and predicted as high as 40,000 species available. About 8,600 are birds' species, mammals 4,500, reptiles' 6,000 species and amphibians 2,500 species.

2.3 Under-utilized Fish

Under-utilized or non-utilized species are those species that offer potential for further processing (Hassall & Associates.1988) or some exploitation in many cases. There are good commercial reasons why a species is under-utilized, such as catching and processing has been attempted or considered, but are simply not profitable at ruling prices.

There are marine species widely used for food and being utilized for industrial or recreational purposes to any great extent and are simply not profitable at ruling prices or risk factor.

2.4 Identity

The fish are mainly classified into five species, the order in class *Chondrichthys* and have cartilaginous skeletons. The *Actinopterygyiian* belong to *Teleostei* and have bony skeletons and ninety-six percent extent fish belong to *teleosts,*

In the past, all animals with so-called "hollow guts" were placed in phylum Coelenterata. Today, scientists recognize two separate phyla as *Cnidaria and Ctenophora* (Coleman, 1987) and *Cuboza* (Edgar, 1997). There are three classes of Phylum *Cnidaria* including Hydrozoa, *Anthozoa* and *Scyphozoa* (Arms and Camp, 1986). Under *Scyphozoa*, the true jellyfish are *Aurelia, Cassiopea, Catostylus, Charybdea, Haliclystus' Pelagia, Periphyla, and Stomolophys,* which consist of larger *medusae* and are more complex and even live longer.

2.4.1 Taxonomy

The classification of fish according to their seven standard categories: kingdom, phylum, class, order, family, genus, and species (Nelson,.J.S.1994). Identification of Many kinds of worldwide fish require concept of grouping formula and arrange into a system according to their variations in the body form as standardization identification for conveniences finding. The following classification of fish base on common characteristic of a notochord;

Phylum:	*Chordata*
Subphylum:	*Vertebrata* (fishes through mammals)
Superclass:	*Gnathostostomata*
Grade:	*Teleostomi*
Class:	*Osteichthyes* (ray-finned fishes)
Subclass:	*Neopterygii*
Division:	*Teleostei*
Subdivision;	*Euteleostei*
Superordder:	*ostariophysi/paracanthopterygii*
Order :	*Perciformes*
Family:	*Priacanthdae*
Genere:	*Priacanthus*
Scientific name:	*Priacantus tayenus* (Richardson, 1846) involved in this project

Source: Fish base, 2003 and Nelson,. J.S. 1994

2.5 General Distribution

The geographical ranges in habituated by fished species. A number of large pelagic fishes such as tuna have circum global distributions throughout tropical and warm temperature waters while some reef fishes such as grunts (*Haemulidae*) are epidemic to single island group such as Galapagos.

The suitable environment for fish population is probably determined by density-independent factors. They form a population and known as stock. The variation is in response to the environment and zooplankton nourishment as food sources for life animal in the sea related to the distribution of the abundance of zooplankton. There is positive relation between the distribution of zooplankton with distribution of demersal fish catches and also the distribution of coastal pelagic catches (Gulland, J.A. 1971)

As mention earliest, the seawater, which covers two-thirds of the earth's surface area where fish seem, have been able to keep pace the development of variety in places of above. At present fish live almost wherever there is water as a medium as well as vital functions of feeding, digestion, assimilation, growth and reproduction as their variety habitat both on the surface and in the surface-connected subterranean waters. They occupy a range from approximately 5 km above sea level to some 11 km beneath as shown in figure 2.2. The whether of waters conditions below freezing to hot spring of more than 40^0 C, from fresh water to salt water in the sea and from sunlit mountain to dark that have never been inhabited by other vertebrates or explored by other mankind. There are approximately 11 300 species found in coastal and littoral water to a depth of 200 meters in the sea. 4500 species are known from coral reefs and 130 marine species have circum global distributions in tropical oceans. Many of these species have very important and high commercial value such as a tuna.

2.6 World Fisheries

Marine fisheries contributed around 80% to the global fisheries or around 90 million tones per year. The growing human population demand more food and improved technology applied in fishing gear to become fisheries industries. The introduction of information technology age in the marketing sector had resulted greater fishing power. Increased competition among fisher or nation has led to the economic collapse of some fisheries. With this scenario the government intervenes to regulate the fisheries.

The humans had fished since prehistoric times especially those who stayed in coastal areas. They had always eaten marine organisms, collected by hand from the shore before invented hooks from bone to catch the fish as found in Greek, and Egypt. It was believed to happen in 8000 BC. The Greek used storage pond and fish farm to ensure a continue supply. Today the fish is being traded or even exported to other countries. With technology support fisheries trade fisheries had became easy where the fisheries industries booming globally.

The European vessels fished across the Atlantic Ocean while the Japanese catch tuna in the sub and Tropical Ocean. The former Union of Soviet Socialist Republics fished for krill (*Euphausia superba*) in the Antarctic in the North Pacific Ocean. During 1970s the catch from USSR, Japan, Spain, Korea and Poland were 7 to 8 million tones and about 10% of world catches.

In highly population area such as South East Asia, Africa and Central America, the fishermen are usually poor. The fish are caught in small scale. The artisan fishermen fished from the shore with small boat with less equipment. They are sufficiently desperate to support themselves and their families and maintained their catch for daily food. Traditional method is applied to catch the fish or even using poison or dynamite although this practice could destroy fish habitat. Their catch has account to 25% of the global catch. (Jenning, S. *et al.* 2001).

The landings increase because increase of fishing effort due to the involvement of high technology in fisheries sector. The investments in this sector have also increased; as a result the catch reduces while some of the individual fisheries collapse. The catch is not profitable because of dominant by catch or under-utilized. Under market size and immature are dominant in most of the catch that only become animal feed or dump to the seas. The cost for handling and processing is too high and uneconomic. When there are competitions to catch large amount of fish among the fishermen the stock will eventually deplete.

Marine scientist had been trying to prevent the depletion in fisheries stock. Good management in fisheries stock was introduced. Stock conservation had been done in many countries such as by biological ways to control the age at first capture or introduce close session to capture in order to allow the fish spawning and recruitments. Introducing regulation to enforce as to monitor the fishing gear and other high selectivity equipment used for catching the fish.

The development of new fishing industries in most countries such as in Southeast Asian for example occurs concurrently by increasing of fishing pressure. A growing number of artisenal fishermen exploits near shore resources. At the same time, the commercial fisheries increase correspondingly and they complies artisenal fisher and commercial fisheries. The complicated and heavily exploited stocks have forced several governments to reassess their fishery development policies (Pauly, D. 1983 (a). The world trade in fisheries industries has grown 116% from 1980 to 1989 (Mohd. Noh, K. 1992). Export grew from US$15.3 billion to US$32.6 billion mostly from developing nations especially Asia. Thus; the fisheries industry is a very important source of foreign currency in Asia.

2.7 Fisheries Technology

Modern technology help to increase fish catch. Rapid developments in fishing activities are correlated with advanced in fishing methods and technology. Development in fishing technology started in 1900 C when several important equipment were installed in fishing vessel to support fishing activities such as hydraulic transmission system (Cunningham *et al*, 1985). Manufacturing industry has played a major part in the creation of new technology. Technology in fishing is also led to increase fishing efficiency and fishing power.

Progressions in capture techniques for example stern trawling stands out as a major development which productivity raising effects when compared with the conventional method of side towing. Moreover, the combined used of son & purse seining and hydraulic power block has been proved spectacularly productive when used in pelagic fisheries. The echo-ranging sonar allows a shoal to be detected: the fish are encircled in the purse seine, a net that can be extended to a depth of several hundred feet and the entire net is then hauled up with power block.

It has been claimed that the increase in total world catch of fish from an estimation 2 million tones a year in the nine tenth century to approximately 75 million tones a year in the early 1980 century was made possible by using mechanical power and electronic aids to harvesting, processing and distribution (Eddie. 1980). On the other hand, increase in levels of technology age are not always accompanied by correspond rises in productivity. The most dramatic Illustration is a case where improved technology has an impact to the catch rates of *Peruvian anchoveta* (notably power block purse seining and the use of synthetic nets). Beside that the herring fisheries in the North Sea and Scandinavia are also an example where improved technology has been accompanied by a reversal in economic performance.

Technological improvements in gear and fishing operation will increase fishing effort and in effective effort may result in over exploitations Innovation in technology also can increase catch and profitability in the short term, but these will decrease in the longer term as the stock is depleted and catch rates are reduced This gradual increase in the efficiency of fishing gear and methods is referred to as technology creep.

The adoption of new technology by fishermen, by raising efficiency and

Profitability, results in intensified fishing by established operators and also acts as an incentive for newcomers to adopt the new technology arid enter the fishery. However, intensification tends to deplete the natural resource which being common property. Adversely affects all users that are exploiting of specifically stock depletion imply a fall in catch per unit of effort. Therefore, the technological progress may contribute to the growth and development of fisheries but may also be indirectly responsible for it's over exploitation.

The gradual increase in the efficiency and methods of fishing gear sometimes referred to as 'technology creep' a double-edged sword. If more fish are being caught in the same amount of apparent effort, profitability in sort term will increase; but if catches continue to increase without control, the fish stock will become over exploited.

A quota was introduced as management tools to control resources. Individual Transferable Quota authorization based on the percentage of Total Allowable Catch (TAC), which will be achieved for long term economic principles.

2.8 The Technological Change in Fishery

Fish harvesting technology has been continuously evolving and this evolution has been especially conspicuous since 1945. Over the past forty yeas, there has been a tendency, for example in many areas fishing has became more capital intensive. The most importance is that the technology has become more sophisticated, as reflected in large number of innovations of past 1945 period, which was originated from fishing and manufacturing sector jointly. This is a pure tic power block, hydraulically powered V-shaped pulley through which large nets can be hauled from the water. The power block was developed by marine construction, which was initially produced a unit for salmon purse seine fleet in the Pacific North-West, and subsequently for many other fisheries.

2.8.1 Fishing Fleet

Development in fishing boats has been invented, designed faster easily driven and increase in efficiency by reducing consumption of fuel. Fiberglass and Ferro cement hulls were used rather woods to extend the economic life of the vessels. A bulbous bow has been implemented in fishery, which significantly improves general vessel performance, and more important, reduces fuel consumption and increase speed of engine power.

Previously the vessel trawled with its gear down with an engine speed of 1400 rpm. However, with the addition of the bulb it was necessary to reduce engine speed to 1300rpm to keep the gear on the bottom. When the vessel maximum speed increases in power there was a corresponding decrease in fuel consumption and estimated to be about 8%.

2.8.2 Navigation Devices

In 1990, a global positioning system (GPS) has become the most important navigational device. Today, the changes in the accumulated numbers of navigational devices can be seen since it is being introduced worldwide. The number of tends are apparent and all point to the fact that fishermen can now increase fishing out of sight their traditional landmarks. This may mean that they are not only able to fish inshore / waters on days when their traditional landmarks are not visible. It is a new accuracy and precision revolution of navigation and fishing operations in shrimp's fishery. By attaching GPS to a plotter, a skipper could record precise tack the boat had taken. Other information can be detected such as position of shrimp mats and suitability of sea bottom.

2.8.3 Fish and Bottom Detection Devices

Color sounders have been introduced to the fishery. Between 1988 and 1981, the use of these sounders had dramatically increased. Most of these devices are mainly employed by the fishermen in herd lining and long lining operations. In 1992, sonar had begun to be used mainly in association with the developing pilchard fishery. Eyesight, light or some aerial spotting normally did the importance of these devices in determining suitable areas and schools of fish.

2.8.4 Knotless Netting

While using knotless netting is not new, improved technology was seen in the construction of stronger material and improved stitch techniques. Knotless netting material is being used to create single layer cod-ends for use in fish trawl fisheries in the North Sea and New Zealand.

2.9 Technological Creep

Effort in fisheries is expressed as nominal and effective effort. Nominal effort is a collective term and is generally derived from the product of vessel numbers and day fished. Effective effort is all the inputs that contribute to the capacity of fishing unit (vessel) to catch. Vessels and all design characteristics, nets, beads electronics, crews, fishing days and hours are all individual units of effective effort. It is common in most fisheries for effective effort levels to change over time. This can occur without a change in nominal effort. In tact, nominal effort will decrease while effective effort continues to rise.

Management targets based on fishing effort, rather than on fishing mortality or catches suffer from the fact that increases in efficiency will cause increases in effective effort even though apparent effort remain the same. This gradual increase in the efficiency of fishing gear and methods is referred to as technology creep. The technology may generally be referred to as the continuous evolution and development in technology in response to the circumstances and conditions that affect how a particular technology can be used to obtain a maximum output.

Technology innovation was established when a given proportion of stocks become more efficient with less cost if the total catch from the stock is limited. By individual catch quota more efficient catching methods has resulted an increased profitability in the fishery. On the other hand, under management systems such as license limitation, where total catch is not restricted. A greater proportion of stock will be taken for the same amount of fishing time. Each unit of fishing effort become more effective and increase in fishing effort may result in over exploitation.

In fishery where catches were un-regulated (open access fishery), the innovations in technology will increase catches and profitability in short term. In the contrary this condition will decrease in longer term as the stock depleted and catch rates reduced. As a result, over fishing have occurred and will decrease the yields. Further increase in fishing effort by increased in technological efficiency would decrease the cost but may eventually reach a point which result in negative return (economic loss) to the fishery. As result depletion occurs in the stock. If there is continuous increase in fishing effort it may drive the stock toward extinction. In this way, technology creep will also poses particular problems in fisheries assessment. Biologists and economist have different point of view about over fishing. Biologist has opinion as "full utilization" of the resource. The fishery can be expanded to Maximum Sustainable Yield and stabilize the biomass. If fishing effort exceeds the MSY, biological over fishing will occur. On the other hand, economist determined that to obtain maximum benefit of harvesting the fishery can be expanded to maximum economic level, where maximum economic level are always in the lower level of MSY for sustainable of stock conservation. If fishery further expands exceed the maximum economic level, economic over fishing will occur and may eventually reach a point which results in negative return (economic loss) to the fishery.

2.10 Management in Fisheries

The main objective of fisheries management has been the conservation of fish stocks. The brood objectives of fisheries management may therefore include the conservation of fisheries resources and their environment. The maximization of economic return from the fishery, and fees payment to the community from profits made by the exploitation of public resources, to ensure sustainable exploitation at the optimum level. This requires controls on the fishery of various kinds such as limiting the number of fishing units, minimum mesh size, closure season and limited entry system. The types of objectives that have been employed can be divided into three groups. Firstly there are those objectives, which are concerned with the attainment of some level of physical yield from the fishery. Secondly, the injection of economics into fisheries research led to consideration of maximum economic yield as a possible objective of management. Thirdly, as a reaction against maximum economic yields the objective of optimum yield has more recently been proposed. This is a concept, with attempts to embody economic, biological, social and political elements into one objective function (Waugh, G. 1964). As mentioned before that economical and biological over fishing is mainly caused the common property nature of fisheries. Without adequate property right, individual's fishermen have less incentive to maintain the fish stock.

It is believed that a number of fishing regulations restricted in the fisheries are sufficient for rational stock management if properly implemented and enforced, There are two types of fisheries regulation in management such as input and output control.

2.10.1 Input Control

Input control is defined as all units that are used in the production process. In fishery ship dimension, crew, fuel and gear are component of input control. Regulations by input control have been implemented in Australian fisheries. These regulations include limiting the efficiency and types of fishing gear range from restrictions on the length of the net or the number of hooks used, to limiting boat Length. Imitations on types of gear may ban specific fishing method such as monofilament gill nets. From strictly economic point of view, gear regulations are in the form of economic inefficiency. Gear regulations raise the cost of catching fish and are therefore reduce efficiency.

2.10.1.1 Limiting the Numbers of Fishing Units

Limited entry licensing has been introduced in fisheries into open access fisheries in many parts of the world. The objective is for license limitation to restrict fishing effort to an amount such as MEY (Maximum Economic Yield) or MSY (Maximum Sustainable Yield). An extension of limited entry licensing was utilized in the mid nineties e.g. Northern Prawn Fisheries and South East trawl fisheries used utilization to control the rise of effective effort.

2.10.1.2 Buy Back Schemes

The buy back schemes are an attempt to remove effort and over capitalization. Buy back schemes have been implemented in South Australia and the Northern Prawn Fishery. However their success in reducing over capitalization and excess effort is unclear when vessels are removed, the remaining vessel tend to catch more fish resulted in increased effective effort.

2.10.1.3 Minimum Mesh Size

Minimum mesh size regulation of nets and trap were used in fishery to allow the escape of small individual and non-targeted species. The method relies on having information on the selectivity of various fishing gear. Increasing mesh size means having to fish in cage to catch the same amount of fish.

2.10.2 Output Control

2.10.2.1 Limiting the Catch

Global quota had been used as a management tool to prevent over fishing. Once the total catch quota is set, the fishermen will compete to secure a large personal catch. When the overall catch quota has reached, fishing activity is closed. The consequences are shorter fishing seasons with increasing costs while some time fishermen have over capitalizes to gain a competitive edge.

2.10.2.2 Individual Transferable Quota (ITQ)

Individual transferable quotas (ITQ) can be allocated to individual fisher. A Total Allowable Catch (TAC) is estimated to the fishery and individual quotas are based on percentage of the TAC. The fishermen have a guarantee share on the resource and have the freedom to increase effective effort within the limit quota allocation. The transferability as one of characteristic regimes gives the fishermen chances not only to exploit resources but also to maintain the future of the stock.

2.10.2.3 Taxes

Taxes can also be an input such as effort for example a levy on fuel would be a tax on effort. Revenue will rise for the management authority and fishermen will use less fuel. It is more appropriate to refer this activity as a royalty, as it is related to the catch.

2.11 The Common Fisheries Policy

As mentioned by Frederick E.Smith later Earl of Birkkenhead in his first edition of his international Law in 1900 at London, that the three-mile limit was generally adopted, and the three-miles rule was a general principle of International Law (Riesenfeld, S.A. 1971). The economic aspects of the policy aimed to modify the Common Fisheries Policy and to improved fishing sector efficiency; the policy was established in 1983.

There are four basic aspects of Common Fisheries Policy namely;

    a.   Limiting catch - 200 miles, open to fishermen community while 12 miles limit to their own shore.

    b.   Marketing organization - covering pricing system and marketing arrangement, the support payment by producer organization was introduced and market discipline was extended to independent fishermen and the Europe Community hold over the prices.

c. Processing and market development projects, conversion and modernization schemes, and redeployment.

d. International negotiations- concerned to waters and conservations of fishing stock since the government has no right to establish its own policy and has no monetary revenues.

Introduction to total allowable catch and quota system is a guide price for fish, multi annual guidance and various subsidies. Basic assumption of this policy was not based on economics but based on political decisions.

This policy was concerned mostly within fishing grounds and quotas. In fact these issues are very difficult to resolve because of the common property resource, which is open to all, and causing to over fishing.

From economist point of view these policy are sub optimal and correct policy would reduce efficiently efforts by massive reduction in capacity of;

a. Paying licenses
b. Tax landing or tax fishing effort

The excess capacity and depletion of fishing stock had little influence to the fishermen who generate highest profit since the price for fishing ground was not applied and enforce. If the basic policy unchanged, permanent problem such as fleet over capacity, the processing industry and over fishing will only accumulate.

2.12 Malaysia Fisheries

The Malaysian location in South East Asian is bordered by marine environments of Andaman Sea to the north and Java Sea to the south. A Malacca strait is on the West Coast and South China Sea on the East Coast of West Malaysia. While Sabah and Sarawak are located in North Borneo also known as East Malaysia separated by South

China Sea in West of Sabah and Sarawak. East Coast. of Sabah facing towards Sulu and Sulawesi Sea. West Malaysia is bounded by Thailand in the north, while in the south Singapore and Indonesia in the west.

Malaysia has tropical climate with summer all the year and humid. The average Malaysia rainfall is 203 cm per year. Malaysia is known as an agriculture country and much of their economic depend on agriculture. However, after 45 years independent, Malaysia has grown as one of the New Industrial countries in the world. In early independent from British colonial, Malaysia succeed in economic planning for became industrial country then the center of information technology in this region.

Apart from information of technology, Malaysia depends in fishing industry as source food and also for exchange foreign currency. Fishery industry had played important role for social economy with (0000 full time fishermen). Fish has become a source for food and animal protein (Consumer Association of Penang, 1977).

The estimated 90,000 population in Malaysia in 1977 were involved in fisheries sector while 3000 to 4000 are yearly unemployed. The fishes landing in 1976 were 375 000 tons compare to 433 000 in previous year or approximately 14% decline. Clear evidence that the fisheries sector is at critical point where 3 out of 4 our fishermen only earn an average income $60.00 per month (Consumer Association of Penang. 1977).

2.12.1 Fish Marketing

Malaysia is among one of new industrial country where fisheries is considered as primary sectors and direct employment to rural and coastal community. Broad fish marketing is divided into two categories related to domestic market and export. It is defined as those commercial activities associated to channeling product from producer to consumer throughout merchandising function to make up the operation function of a business. Most of the fish catches landed at privately owned landing complexes or jetties.

Fish will be transported from producing to consuming areas is fairly well organized; fishes are packed in ice cube and transported in wooden boxes and collected by agents at producing areas. Fish goes throughout a number of stages; the wholesalers bought the fish at the production area. This fish then divided to small quantity for sell to the final consumer. At this stage the wholesalers play a major role in determining the fish prices and the quantities to be purchased and then when to sell consumer. The wholesalers maintain daily product and adjusted the prices volume the quantities for sell.

Fisheries in Malaysia play an important role and contribution to the economic nation as food supply, providing job to the people and earning foreign exchange. Fish has come to be the cheapest source of protein food and approximately 80% of the total protein intake from animal. There is a relatively high average per capita consumption of about 40 kg. Per. Annum (Jomo, K.S. 1991).

As a renewable resource and with proper management this resource could be sustainable. Unfortunately over fishing occur towards these resources where a lot of fish were removed from the seas leaving behind little resource for further recruitment. Thus when fishing mortality occur too high has resulted decline of the resource and finally parental stock cannot be recovered and declining and over fishing is going on (Consumer Association of Penang, 1977).

2.13 *Priacanthidae* Fishery

The family *Priacanthidae* contains four genera and four species that occur in western central North Atlantic. *Pristigenys alta* is distributed in the Caribbean, Gulf of Mexico and along the east coast of North America. Although juveniles have been reported from as far north as southern New England waters, adults are not reported north of Cape Hatteras, North Caribbean. *Priacanthus arenatus* is distributed in tropical and tropically influenced areas of the western central North Atlantic in insular and continental shelf waters.

2.13.1 Taxonomy identification

Kingdom *Animalia*

Phylum: *Chordata*

Class: *Osteichthyes (bony fishes)*

Order: *Perciformes* (perches)

Family: *Priacanthidae* (bigeyes)

Habitat: Marine

2.13.2 Distribution

Adult *P. arenatus* are distributed north to North Carolina and Bermuda, juveniles have been collected as far north as Nova Scotia. *Cookeolus japonicus* and *Heteropriacanthus cruentatus* are circumglobally distributed species and are both common in insular habitats. In western central of North Atlantic, *C.japonicusranges* from New Jersey to Argentina; *H. cruentatus* from New Jersey and northern Gulf of Mexico to southern Brazil. Adult *priacanthids* are small to medium size fish that are characterized by large eyes, which have a brilliant reflective layer, extremely rough scales and bright red live coloration.

They are epibenthic predatory fishes that inhabit primarily rocky or coral habitats in depths from 5 to at least 400 m. *Priacanthids* in the western central North Atlantic appear to spawn in summer and fall; *P. alta*: late June mid September; *C. japonicus:* May-September in the Caribbean; *H. cruentatus:* late spring-early summer in the Caribbean; and *P. arenatus:* fall in the Caribbean. *Priacanthid* eggs are pelagic, but egg size and hatching size are unknown. Flexion occurs in 4-5 mm NL and the dorsal and anal rays are completed at approximately 7 mm SL. Larval *priacanthids* have a large head that has well developed spination consisting of a series of parallel serrated ridges on the dorsal surface of the frontals, and a large vaulted, serrated crest that is well developed in preflexion larvae soon after hatching. Other head spination includes serrate preopercle marginal spines, opercle, subopercle, interopercle, tabular, posttemporal, supracleithrum, lacrimal, circumorbital, nasal, dentary and branchiostegal spines. Spine lengths attain their greatest relative size during the preflexion stage and the diagnostic supraoccipital crest and spine disappears in the pelagic juvenile stage (approximately 30 mm TL). *Priacanthid* larvae are heavily pigmented in the head and gut region and with development on the lateral surfaces of the body.

Priacanthids have a pelagic juvenile stage that is usually barred and mottled and undergoes considerable change in appearance during settlement to their demersal habitat. Larval priacanthids are distinctive in both pigmentation and spination and are not likely to be confused with any other larvae in the western central North Atlantic. They may superficially resemble holocentrids or scombropids, which exhibit a supraoccipital crest and spine, but if confused they can be separated from these taxa by myomere counts (*scombropids* and *holocentrids* have 26 vertebrae, *priacanthids* have 23).

Furthermore larval *holocentrids* attain a characteristic rostral spine that is apparent at approximately 2.8 mm TL. The only larval *priacanthids* identified from North Carolina waters (from meristic characters) were *P. alta*. Although Powell were unable to confirm the identification of larvae that had not developed meristic characters, it is likely that the small larvae were dominated by *P. alta*. Observations (Powell A. B. 2003)

2.13.3 Feature

Body:	Oblong, very compressed.
Head:	Eye large, slight more than snout; mouth strongly oblique.
Dorsal fin:	X spines, 13 to 14 rays.
Anal fin:	X spines, 13 to 15 rays.
Pelvic fins:	Larger than pectoral fins, reaching onto anus.
Caudal fin:	Early truncate.
Scales:	Lateral scales 68 to 83.
Color:	Body vermilion red, ventral lighter color and numerous olive or brown spots on dorsal, anal and pelvic fins.
Size:	Attains 40cm total length, common size unknown, specimen size 15cm TL.
Distribution:	East Indo-West Pacific: from southern Japan in the north to western Indonesia, the Arafura Sea and Australia in the south.
Habitat:	On inshore or offshore reefs from less than 20 m to more than 400 m depth.
Economy:	Common food fish with fairly good price. By capture fishery, rarely Cooking Method: Steam or soup.
Environment:	Marine

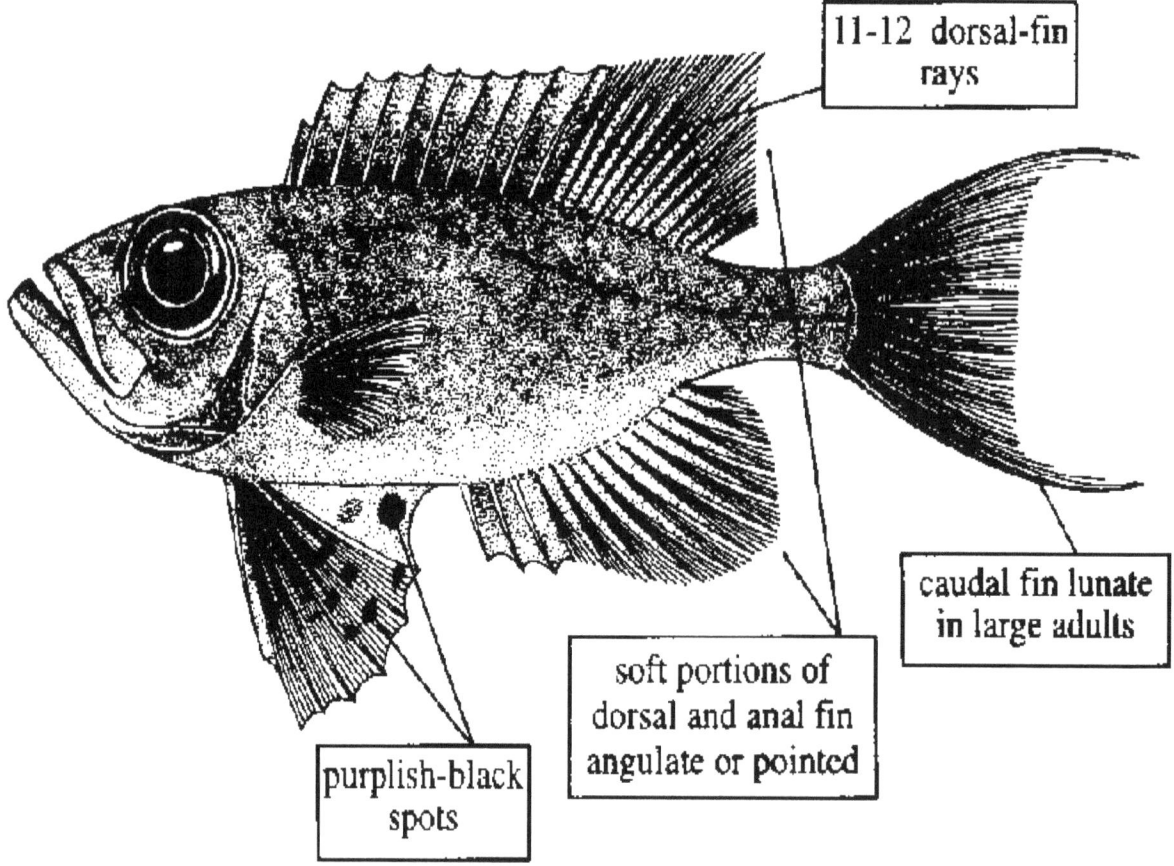

Source: fishbase, 2003

Figure 2.1 Anatomy of *Priacanthidae*

2.14 Threadfin big eye or *Priacanthus tayenus* (Richardson, 1846)

The species can be found in tropical and subtropical Atlantic, Indian, and Pacific oceans. The eyes are very big, with a brilliant reflective layer (tapetum lucidum) with and superior (strongly oblique) big mouth. The dorsal fin spines are usually 10; soft rays 10-15. Three spines are found in anal fin; soft rays 9-16. The caudal fin slightly emarginated to round. There are sixteen principal rays in caudal fin (2 unbranched). Inner rays of pelvic fin attached to body by a membrane. The scales are very rough with integral spines, usually bright red in color. It is Epibenthic and generally associated with rock formations or coral reefs; a few species are often trawled in more open areas; usually carnivorous and nocturnal. The eggs, larvae and early juvenile stages are pelagic. Typically less than 30 cm TL, but the largest species attains more than 50 cm maximum length. The species white meat is considered as superior meat and has economy value for human food.

Habitat in shallow waters to depths from 150 to 200 m over coral and rocky formations, occasionally form dense aggregations.

2.14.1 Identification

Family: *Priacanthidae*

Order: *Perciformes*

Class: *Actinopterygii* (ray-finned fishes)

2.14.2   Distribution

The species is distributed in Indo-West Pacific: from the Persian Gulf to the western coast of India, eastward to the Pacific, where it occurs from Taiwan Island southward to the Arafura Sea and northern Queensland, Australia.

2.14.3   Biology

The species biologically Inhabits coral or rocky areas and sometimes form aggregations. Feeds on a wide variety of benthic animals. Smaller fish are mostly found inshore. Marketed mostly fresh, whole or dried-salted.

2.14.4 Appearances

Body:	Elliptical and compressed
Eye:	Very large
Mouth:	Large
Dorsal fin:	Continuous, X spines, 11 to 15 rays
Anal fin:	III spines, 10 to 16 rays, soft parts of dorsal and anal fins angulated or pointed.
Caudal fin:	16 principal rays, slightly emarginated to rounded, lunatic, often with both upper and lower filaments.

Pelvic fins:	Membrane present connecting the inner rays to the body and shorter than head, joined to body by membrane
Scales:	distribute on branchiostegal membrane, rough, difficult to detach.
Vertebrae:	23
Color:	Body brilliant crimson red usually bright red, paler below. Pelvic fins with distinct blackish red spots, other fins without spots.
Commercial importance:	Common commercial food fish
Max. size:	35.0 cm TL (male/unsexed)

Environment: reef-associated; Marine; depth range 20 - 200 m

Climate: tropical; Range distribution 30°N - 19°S

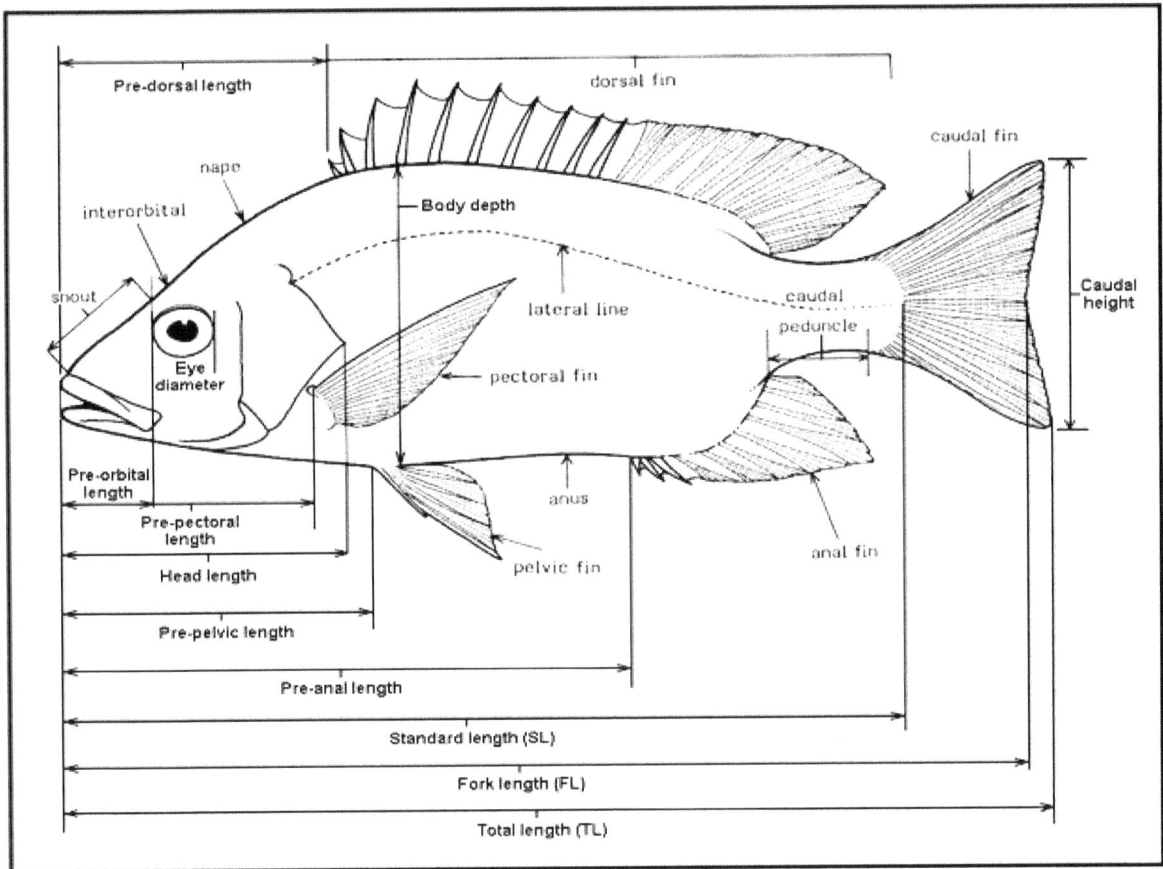

External Morphology and Measurements.

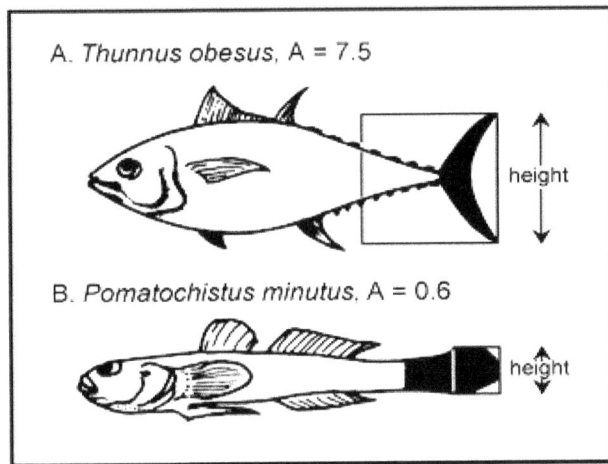

Aspect ratio ($A = h^2/s$, h = height of the caudal fin; s = surface area of fin) of a pelagic fish ($A = 7.5$) and a bottom dweller ($B = 0.6$). Note the correspondence between aspect ratios and modes of life.

Source: Fish base, 2003

Figure 2.2 External morphology and measurement

Source: fish base, 2003

Figure 2.3 The Threadfin big eye (*Priacanthus tayenus*)

2.14.5 Morphometrics

Size (cm): 18.7 SL

Standard length: 85.9% TL

Fork length 100% TL

Pre-anal length: 44.6% TL

Pre-dorsal length: 22.4% TL

Pre-pelvic length: 17.8% TL

Pre-pectoral length: 22.6% TL

Body depth: 30.2% TL

Head length (HL): 22.4% TL

Eye diameter: 42.7% HL

Pre-orbital length: 25.6% HL

Aspect ratio of caudal fin: 2.14

2.15 Malaysian Fisheries

The fisheries sector plays an important role in the national economy in terms of income, gain of foreign exchange and employment to the rural population.

Fishing industry is managed by strict input restriction management regime such as fishing licenses, boat and engine restriction, area limitation and closed area for traditional fishermen. As the area operation is also shipping routes, safety at sea is given highest priority in the safety of the vessels. Aquaculture is given such as priority for future expansion and more coastal area is being identified for the purpose. Fishermen are advised to shift toward aquaculture to ease conserve on the marine resources. Malaysia Fisheries policy has been direct at alleviating the poverty of its fishermen. Efforts have been made to reduce the number of people dependent on the resources. The main concern is to reduce over fishing within 30 nautical mile of the coast along the bay and estuarine fisheries of most rivers are well developed especially in dense populated area.

Basically the fisheries industry in Malaysia can be divided into marine fisheries, aquaculture and public water bodies/inland -fisheries. However, a public water bodies/inland fishery at present does not contribute significantlyto the fisheries industry. For management purposes, the marine fisheries is categorized into inshore sector and deep sea sector. Aquaculture on the other hand covers freshwater culture and brackishfishwater culture. The inshore fishery is already heavily exploited and there is evidence that fishing is over saturated. However, the deep sea fisheries have further potential for development. Aquaculture also offers bright prospects for further development and expansion. Based on resource potential, the developmental efforts are geared towards the deep sea fisheries and aquaculture sector while sustaining production from the inshore sector. Ornamental fish industry in the past few years had shown significant growth in terms of production and export.

The various ecosystem in Malaysia such as estuaries, mangrove swamps, coastal areas, coral reef and continental shelves are suitable for fisheries industries. The coastal line along 3,800 km occupied by coastal community where most of them are fishermen. The coastal area is a source of food for the coastal community to catch fish and other seafood as their primer food. During the northeasterly and southeasterly, the coastal area are subjected to wave and form an extensive coastal plains with dominant deposition of fine grained sediment. These interactions have formed mud flat, swamp and fertile productivity of marine resources as food sources for the coastal community. The fish and prawn both have contributed as food sources and also social economy for most of the coastal community. At the same time it has became an economy strengthen to the country for domestic consumption and export commodity to gain foreign exchange.

Mangrove plays an important role in marine environment ecosystem for stabilizing along the coastline. Beside that, mangrove is also acting as a retainer and form of land, to protect the coastal area against waves and storms (hurricanes and cyclones), a protector of beaches, stream banks and a reservoir in the tertiary could assimilation the wastes. Chemically, mangrove has an importance role to form and source of inorganic for plankton production. Biologically, mangrove forests in the coastal and tidal area has formed habitat for fish and shrimp (Arshad *et al.*, 1997). Mangrove can trap the sediment thought pneumatophores and fine roots. Mangroves forest is also as habitat not only the fish and prawn but also other organism and invertebrate in the estuarine and near the shore waters for food sources. Apart from fish habitat, mangrove can also produce high abundant of plant material, as well as food source for the estuarine habitat and coastal community.

Fisheries industries plays an important role in the Malaysian economic to produce seafood can affect imports commodity. Seafood is always available in Malaysia and had become the first and popular food for most people.. Seafood is consumed during breakfast, lunch and dinner or even supper. In the evening snack such as keropok lekor has gained market place and came to be the first choice for most Malaysian especially in Terangganu and Kelantan.

The supply of seafood in Malaysia is about seventy percent of the intake of total animal protein. Malaysian seafood contains high nutrition value, safe to be consumed and can be found in most retailer and supermarket at the cheapest prize. Most of he coastal community involve in fishing activity because of their strategic locality. The village in the coastal area had made the community adopted fishermen as their way of live. This community form and become artisenal fishery where simple fishing gear by using small boat of low power engine or without engine.

The government had built infrastructures and designed other facility for fisheries activity such as fish landing or jetty. The infrastructure provided for fish landing as well as for marketing purpose. Fish market is built to upgrade the social status of poor fisherman.

The fisheries in Malaysia are considered as common property resources. Therefore efficient management is greatly needed to control marine resources for sustainable purpose in future. There are several Fisheries regulations of Malaysia and Federal Legislations made under Fisheries Act 1963 and various subsidiary legislations have been revised on 1978, then Fisheries Act 1984 with revised on 1986 the other are: -

- a. Fisheries (Cockles Conservation and Culture) Regulations, 1964 L.N 428/1964.
- b. Fisheries (Maritime) Regulations, 1967 – P.U. 49/1967.
- c. Fisheries (Prohibition of Method of Fishing) Regulations, 1971 – P.U. (A) 187/1971.
- d. Fisheries (Prohibition of Import, etc. of "piranhas") Regulations, 1973 – P.A 355/1973.
- e. Fisheries (Maritime) (Sarawak) Regulations, 1976 (P.U. (A) 401/1976).
- f. Fisheries (Prohibition of Method of Fishing) Regulations, 1971 P.U. (A) 187 of 1971

2.16 Sabah Fisheries

Fisheries sector are important and popular for economic development in every developed or developing country. Fisheries have become food source and income to the society in the coastal state. The activities in fisheries had contributed income to fishermen. However, in the other aspect the fish stock are being exploited beyond Maximum Sustainable Yeld (MSY) or over exploited and over capitalized where revenue can not cover the cost and as a result the fishermen lost their profit. At this level the manager should include a programme of recruitment to avoid stock depletion, as the stock will not be recovered by recruitment of stock.

This phenomenon is needed to solve the problem being expanding and serious. The manager should protect the stock being over exploited by reducing by catch or reducing too many people depended on fisheries on their live hoods. This situation has lead the fishing industries economically marginal as the stock is being depleted. Environmental damage occurs because of fishing technologist and unhealthy fishing practice such as using dynamite and cyanide poison. In future global conservation standard and guideline focusing on long term are needed for all fisheries management plans.

The management in developing country is relative difficult. The fisheries manager do not have enough data for assessment and pressure often arise from politician who propaganda the resources to gain vote in the election. As a result the industry has become extremely high of competition to catch the fishes.

The Fisheries Department of Sabah, is responsible to manage the fisheries stock in the state. Situated in Northern Borneo and better known as East Malaysia has 1.5 million populations with 5 percent full time or part time fishermen.

The Fisheries Department is also responsible to manage living marine resources with specific objectives and goals. They include conservation and protection to allocate the resources between user and other state holder and to ensure the integration of environmental and economic consideration to encourage the optimum use of the resource without effect of decreasing its long run productive potential. The stock decline may not because of level of exploitation but due to the environmental deterioration. Many users threaten the marine environmental by using physical and chemical. Using dynamite and cyanide as a method to get the fish instantly by some fishermen may cause destruction to the species and the coral reef.

Fisheries as a whole suffer from both commercial and traditional over fishing. The dilemma is that as the demands of fisheries resources increases; the ability of the marine environmental to sustain them may be decreasing. Thus integrated coastal management and Ecological Sustainable Development are much needed to ensure the sustainability of resources. This is challenge to Sabah Fisheries Department as well fisheries manager to regulate exploitation prior to over harvest and must be consistent with the aims of maximizing management of the fishes resources and minimizing logistical and technical limitation.

2.16.1 Research and Development

The research division is established to focus on long term program of fisheries management. The researcher banding into fisheries management process as an advisory to the management level where the management strategies to establish the status of resource and to determine the level of exploitation. An addition of biological information, environmental, socioeconomic and political consideration has to be viewed by all managers. The problem to the fisheries was lack of human resources such as Fisheries Scientist to manage the resources to provide biological data in the State.

2.16.2 Management Regulation

The resources conservations in Sabah were monitor by 16 marine districts. These districts involve in enforcement of current regulation as contained in Fisheries Act, 1985. The consternation proposed including restriction of fishing effort and age at first capture.

2.16.3 Input Control

2.16.3.1 Limiting the Efficiency and Type of Fishing Gear

This limit to the number hooks used. Most regulation sticks to 2-3 hooks per line. Control on the number of rod/line is also being in forced but insignificant as the regulation on bag limit compensates this.

Specific gears such as trap need to be controlled and restriction should be made on license and specific size and their application. Net and other gear that are similar to commercial fishing but smaller size should not be encouraged.

2.16.3.2 Limiting the Number of Angler and Fishing Unit

The limited entry aims to restrict fishing effort and hence pressure to the resources. The number allowed may depend on the Catch information per unit effort. Number of fishing gear should be restricted whether the level of exploitation needs to be reduced or maintained. Limiting entry of user may rise to series conflicts between the government and the user. Implementation of license regime should be considered.

2.16.4 Output Control

2.16.4.1 Size Limit catch

The main aim of size limitation is to maintain adequate breeding stock and promotion of maximum yield, which is regarded, as desirable. The limiting factors may be on size and weight of the catches. The aim of size limitation is to protect spawning stock and produce at least 50 percent changes of breeding at least once.

2.16.4.2 Rejection of Spawning Female or Female

Making it an illegal to retain female or female bearing eggs is one method of conservation. However, fast growing and high fecundity species is ineffective as a study had shown that breeding stocks can be removed without affecting the stock. Protection of female can only apply if sexes can be identifying by naked eye.

2.16.4.3 Total Protection

Heavily exploited or near extinction species should be totally protected until stock recovered. Heavy fines or juridical proceeding taken to those engage in fishing the species.

2.16.4.4 Zoning

The zoning to ensuring equitable distribution of income and alleviating the poverty, that exists for many of its inshore traditional fishermen. This method of regulation used to separate potentially conflicting resources users such as the difference of gear used. Zoning to identify traditional area and commercial area may be considered especially in closed area as lagoon, along the coast, river, lakes and dam.

2.17 Darvel Bay

The Darvel Bay in South East Asia as the traffic of vessel for trade or fishing activity. Its water which boarding by three countries namely Malaysia, Philippine, and Indonesia, has played a variety of rules, contribute not only to the border country but also to the international as well. The main roles of the bay are as an international and domestic waterway, a major fishing ground and an aquaculture area, a coastal tourism.

The Darvel Bay routes mainly used for cargo shipping to carry trade especially for the BIMP- EAGA members (Malaysia, Indonesia and Philippine). This bay is also as the international shipping, bound trade especially tanker with crude oil palm and petroleum vessel, large proportion being tanks. There will be threats the marine animal at seawater or at aquaculture project. The conflict occurs between users such as fishermen, government and public, conflict between fishing vessels and commercial vessels.

The pollution is a bad thing, but for scientific need to value judgments of this kind to be qualified of the bad. Pollution commonly mean the environmental damage caused by wastes, discharged into the sea, the accuracy of wastes in the sea and the waster themselves. Marine pollution is introduced by man, directly or indirectly and causes in deleterious effect such as hazards to human health as well marine habitats. The high concentration substance in the sea may cause the death of the same plants and animals in the natural environmental.

2.17.1 Coastal Tourism

Tourism is becoming a lucrative activity. There is several ecotourism resorts along the coastal area. Seawater recreations and sport activities are also becoming popular events along the coastal water in the bay. The present of coral reef and sandy beaches along the coastal bay are promoting the growth of tourism in the area.

2.17.2 Coastal Aquaculture

Basically, the aquaculture industries are more practice in this bay comprises of freshwater and brackish water culture. Freshwater culture is carried out in freshwater ponds, freshwater cages. On the other hand, brackish water culture covers brackish water ponds, brackish water cages and mollusks culture comprising of cockle, mussel and oyster culture.

The coastal aquaculture is becoming an important alternative activity in the area due to depletion in landings of marine catches. These involve cage cultures in the coastal water and brackish water ponds along the inter tidal zone. Huge areas of mangrove forest along the coastal area have been cleared and converted to the brackish water.

2.17.3 Pollution

Malaysia has a responsibility to prevent damage to marine environment, resources and also to protect neighbor's resources being polluted from activities under its jurisdictions. Nevertheless from the scenario mention above we have seen that various specific department and ministries under specific area of jurisdiction manage the various users. Beside the federal government and state government have a different opinion on some overlapping area and concurrent issues. Thus, there is a weak unilateral link in management and legal aspect pertaining to marine uses, protection and conservation management regime. As a result similar issues addressed by various departments and some time finger pointing for action and responsibilities.

2.17.4 Illegal Fishing

It is difficult to enforce the regulation along the coastal area from Tungku to Pulau Timbun Mata near Semporna district because of limited personal. Instant catch by under water explosion or by bomb could destroy the habitat and also juveniles of fish. This activity often occurs in Malaysia especially in Sabah water and uncontrollable because some of the bomber are full time fisherman. This is happened due to lack of cooperation between the fisherman, fishmonger and public. In the nineties, the explosive bomb could be heard almost every day in this area. It was dangerous not only to marine resources but also to fishermen or to the scuba swim in this area.

2.17.5 Coral Reefs

The reefs within Darvel Bay have been extensively damage by hand made explosive fishing gear. However, isolated patches remain with high coral cover and fish diversity. The outer reefs, further from land and out of the normal range of local fishermen, are a mixture of good and poor condition reefs. The distance from shore has positive and negative benefits of the reefs. Lack of enforcement attracts a few blast fishermen while majorities remain closer to shore. Many reefs in this region have large numbers of *Diadema* urchins, indicating an imbalance in fish diversity with the removal of the normal grazing community.

The percentage of coral cover in Lahad Datu and Silam in Darvel Bay spanned a broad range. There hard coral covers more than 75% but other reefs had live coral cover of less than 30%. Darvel Bay is dominant with hard corals of 225 species (67 genera; 15 families) identified (Pilcher, N., and Cabanban, A. 2003).

Recovery of coral communities on reef areas following major impacts, such as coral bleaching, requires the settlement and growth of new larvae and asexual reproduction of remaining colonies. Although these were relatively healthy areas, it seems reasonable to conclude that there are viable coral larvae in the water column and that they will also settle on damaged reefs. In addition to the presence of young corals, the vitality of the area was demonstrated by the observation of a mass coral spawning.

2.17.6 Oil Pollution

Oil pollution in the sea caused when petroleum hydrocarbons reach the sea by waterways and maritime activity. The probability pollution by oil spills Petroleum hydrocarbons reach the sea by many routes, but the tanker operation major contribution to the oil spill at the bay.

The production of crude of palm oil is highly transported by sea. The ballast water stored in the cargo compartment, which previously contained oil, will mix with oil remain clinging to the wall of the compartment and this dirty ballast water were discharged to the sea. While docking all the cargo including the oil need to be transferred to avoid the risk of explosion from petroleum gaseous. During their transfer the oil may reach the sea.

2.17.7 Damage from Pollution

Insecticides, pesticides, floating oil and waste of fertilizers from agriculture plantation will affect the adult fish. The fish egg and larvae is more sensitive than adult to toxin. The concentration of hydrocarbon on water surface where crude oil accumulated reduced the hatching success of fertilized egg. It may also cause abnormalities development in several species. Extensive mortality of marine organism may occur because of oil spills.

Marine and coastal pollution has a negative impact on coastal ecology and causes economic and financial losses as for clean-up operation. The ecological damage cause by such pollution has to be fully determined. Annually the government must allocate substantial funds to dredge millions of tones of sediment from harbors and navigational.

2.17.8 Management

Beginning of UNCLOS, AGENDA 21, etc. embellished awareness of marine multiple uses parallel with initiative at international level such as UNCED, IMO, WWF, MAARPOL, Greens and Malaysian Fisheries Act etc. with new and integrated management approaches. Law and jurisdiction must be redesign for conservation and control of marine resources. There is provision concerning compliance of law with heavy fines for illegal fishing at the strait.

Management and jurisdiction of multiple uses in the strait bay been established since the British's colonialization. Various law, jurisdiction and management regime such as Merchant shipping ordinance (1952) and the Merchant shipping (oil pollution) Act 1993, Environment Act, Fisheries Act 1985, The state inland and river enactments, state enactment etc. Thus there is a conflict of responsibilities that some still happen till to this date such as land development and riverine is a state matter and matter pertaining to sea is a Federal matter that eventually resulted in complexity in enforcement social and economic development toward sustainable resources.

2.17.9 The Conflict of Uses

At the same time the bay marine water quality is significantly affected by multifarious lad use activities found in this area.

The landing of fish may influence by direct or indirect way reducing of biomass because of mass mortality of parent stock, decreased catches may also occur because of effects of pollution the species with free swimming may migration ration to other place where they fell safety, the fishing gear are frequently clogged by crude oil the trawl caught the oil at bottom sea bed. The colours of the marine animal also modified the consumer not favour with the product and cause the product low of demand and worthless in the marketing then the fishermen will lost their profit.

The level of conflict among uses in bay are international waterway, fishing ground, coastal aquacultures, coastal tourism with land use activities The major treated is the oil spill from the vessel users the resulted in inevitable sea based pollution. In addition water ballast may also affect the marine habitat. Deforestation for aquaculture cause sedimentation problem to the ports, reducing recruitment in the marine fisheries and erosion problem along the coastal area, on ones orientation can be seen playing a variety of roles, a major fishing ground an aquaculture area a marine resources or a coastal tourism belt.

The Environment Quality Act.1974 with regard to the marine pollution as well as difficulty in enforcement with reference to non-point sources of pollution, the sea bed and its sub soil are not specifically included in its definition at the Act. Deforestation and utilization of coastal forest and mangrove for urbanization, industrial, shipping and infrastructure would damage to its marine environmental. The conflict between artisanal fishing with lowest effort to catch the fishes and commercial fisher whose with high investments by using high technology fishing vessel with high effort to catch the fish this happen in Tungku sub district where the restricted area at 5 fathom line for all type of trawler.

2.17.10 Fisheries Status

2.17.10.1 Marine Fishes Landing

The catch landing where the data available from 2 marine districts in this bay shows that during 1991 to 2000 the average marine fish landing was estimated 31654.179 tones. In 2000, the total production from the fisheries sector amounted to 53051.36 tonnes. There was an increase of 39.5% in terms of the quantity of the total production from The fishes landing from traditional vessels and commercial vessels operating all types of gears The estimated of annually marine fish landings during 1991 to 2000 as Table 2.1 and Figure 2.4.

Table 2.1 Marine fishes landing in Darvel Bay

Year	Quantity (Tones)
1991	13761
1992	21092
1993	29032
1994	29957
1995	32092
1996	40161.36
1997	28317.45
1998	31048.78
1999	38028.84
2000	53051.36

Source: Sabah Fisheries Department Annually Report, 1992, 1993, 1994, 1995, 1996, 1997, 1998, 1999 and 2000)

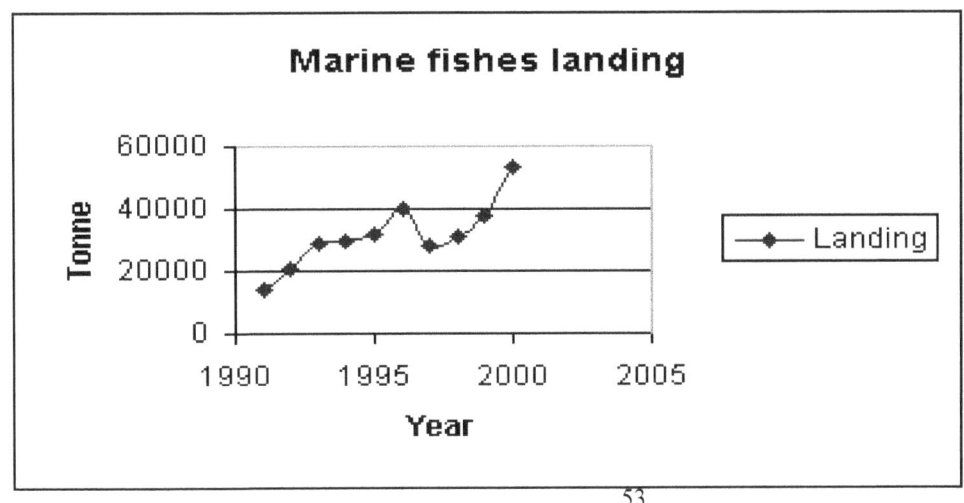

Figure 2.4 Marine fishes landing in Darvel Bay

2.17.10.2 Prawn Landing

In 1999, There were about 562 tones prawn landing in two districts of Lahad Datu and Kunak the gear catch by trawlers the trawlers were owned and operated by small-scale fishermen. There was an increase of 160.2% in terms of the quantity of the total production from The fishes landing from traditional vessels and commercial vessels operating all types of gears The estimated of annually the landing prawn catch during 1991 to 2000 are shown in Table 2.2 and Figure 2.5 below:

Table 2.2 The prawn landing (tones)

Year	Quantity (tones)
1991	78
1992	134
1993	160
1994	212
1995	138
1996	114
1997	154
1998	216
1999	562

Source: Sabah Fisheries Department Annually Report, 1992, 1993, 1994, 1995, 1996, 1997, 1998, 1999 and 2000)

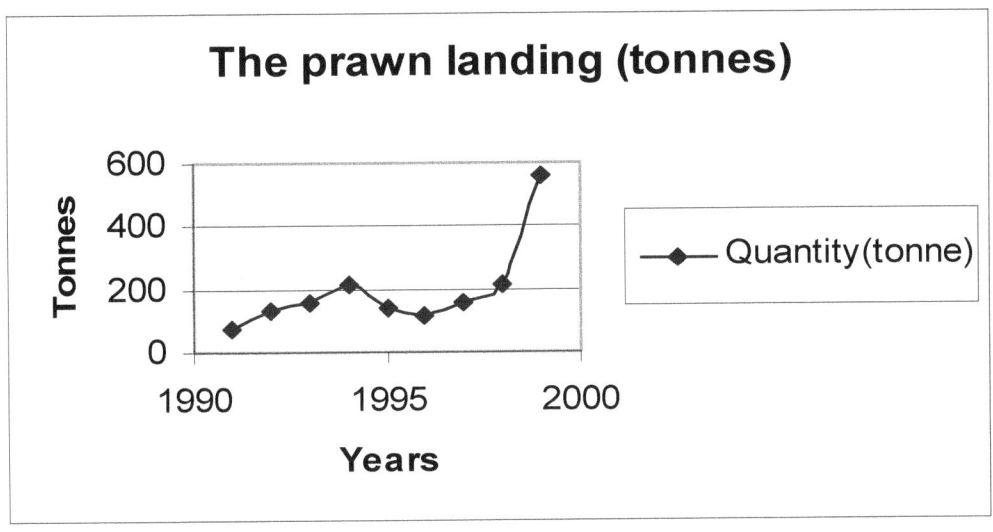

Figure 2.5 The prawn landing (tones)

2.17.10.3 Value

There was an increase of 58.4% in terms of of the value of the total production the total marine fish landing The estimated of annually the value production during 1991 to 2000 are shown in Table 2.3 and Figure 2.6 below:

.Table 2.3 The marine fishes value (RM000)

Year	Value ($000)
1991	13957.97
1992	19776.7
1993	38834.29
1994	40426.87
1995	50672.32
1996	55114.73
1997	54679.18
1998	65104.97
1999	82678.14
2000	130975.95

Source: Sabah Fisheries Department Annually Report, 1992, 1993, 1994, 1995, 1996, 1997, 1998, 1999 and 2000)

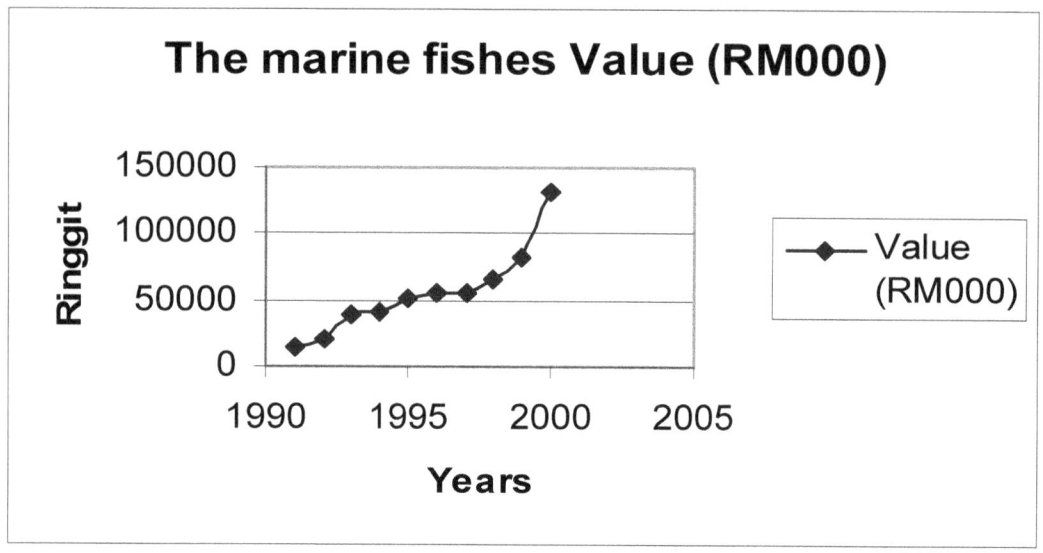

Figure 2.6 The marine fishes value (RM000)

2.17.10.4 Employment

The fishermen in Sabah can be divided into two categories, the commercial fishing operation with trawl net, seining nets, and gill net, which is, better organize. More capital investments hunt for greater income and the other the traditional fisheries by using various traditional gears such as trap, hook and line, liftnets and bagang. Bagang is origin Indonesian traditional fishing gear with small investments and highly intensive labor and smallest income or as a source food for the coastal people.

A total of 1470 fishermen were engaged in marine fisheries sector in 1998 compared to 1470 fishermen in 1999. This shows that there was no decrease or increase of fishermen in these two districts. This trend is in line with the Government management policy to reduce the number of fishermen so that each one will have a bigger share of the resources. Only the fishermen who are genuinely interested in fishing as a means of livelihood will stay in the industry. The estimated of fishermen in 1990 to 1999 are shown in Table 2.4 and Figure 2.7 below:

Table 2.4 The number of fishermen

Year	Number
1990	700
1991	831
1992	997
1993	1088
1994	1203
1995	1203
1996	1227
1997	1227
1998	1470
1999	1470

Source: Sabah Fisheries Department Annually Report, 1992, 1993, 1994, 1995, 1996, 1997, 1998, 1999 and 2000)

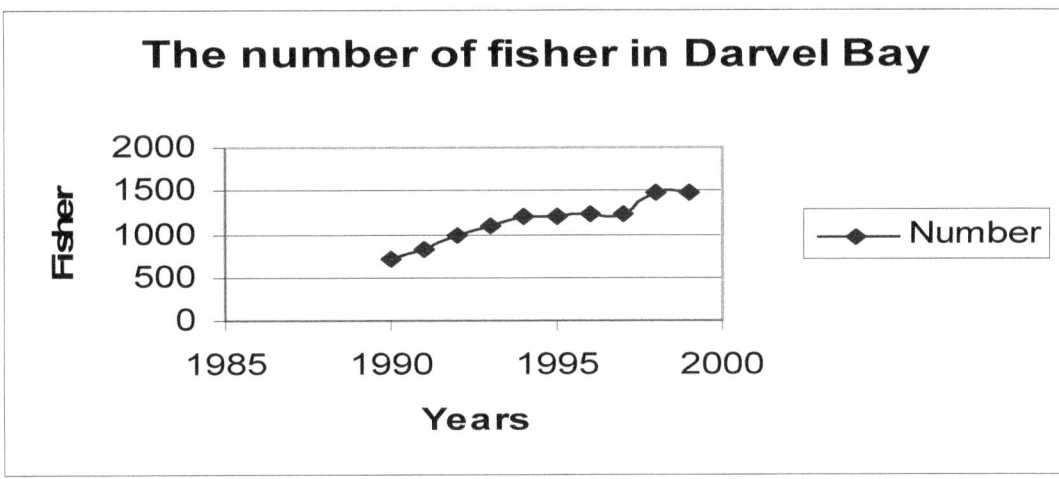

Figure 2.7 The number of fishermen

2.17.10.5 Fishing Gear

The total number of fishing gears licensed in 1999 and 1998 were constant at 795 units. There are no increasing or decreasing of fishing gear during 1998 and 1999. The trawler is constant with 31, 30 purse seining and 734 respectively belong to traditional. The sum of all together fishing gears were 795. Table 2.5 and Figure 2.8 show the numbers of fishing gear licened in this bay.

Table 2.5 The number of commercial fishing gear

Year	Trawler	Purse seine	Other	Total

1990	10	23	249	282
1991	36	22	333	391
1992	47	26	402	475
1993	47	28	414	489
1994	37	26	518	581
1995	37	26	373	436
1996	16	27	623	666
1997	16	27	623	666
1998	31	30	734	795
1999	31	30	734	795

Source: Sabah Fisheries Department Annually Report, 1992, 1993, 1994, 1995, 1996, 1997, 1998, 1999 and 2000)

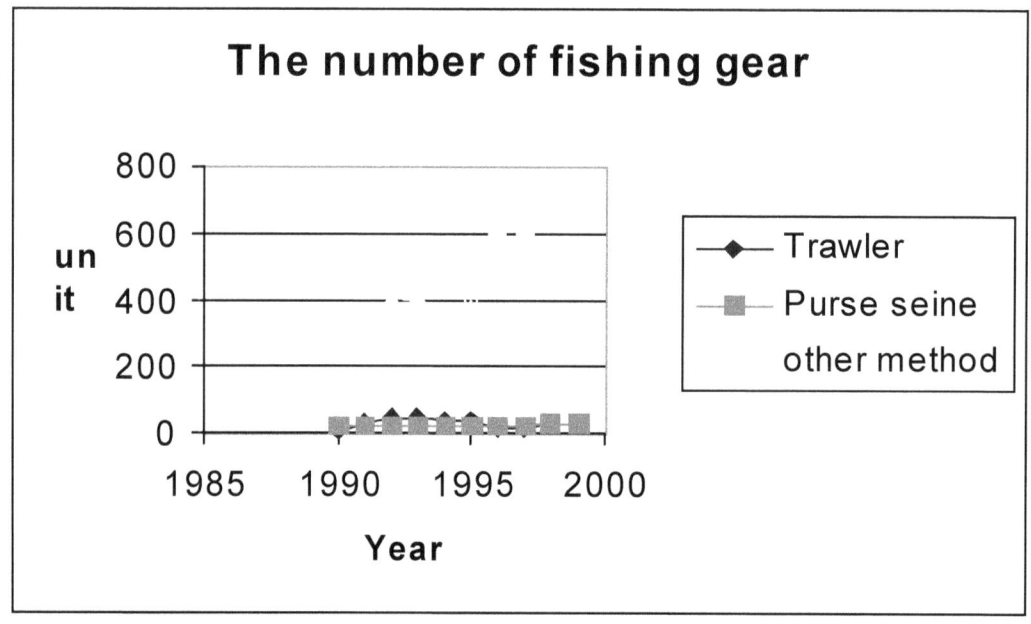

Figure 2.8 The number of commercial fishing gear

2.17.10.6 Fishing Boats

The variety fishing boat can be divided into three types. Fishing boats operated by type of engine as, in boat engine (158), out boat engine (202) and boat without engine (202). In 1999 and 1998 there was no increasing or decreasing in three engine types of boats in the last two years. Table 2.6 and Figure 2.9 show the numbers of fishing boat according to their engine type.

Table 2.6 The numbers of fishing boat according to the engine type

Year	In boat	Out boat	Without engine
1991	140	116	94
1992	138	172	165
1993	140	181	168
1994	163	121	360
1995	163	121	152
1996	143	126	187
1997	143	126	187
1998	158	202	202
1999	158	202	202

Source: Sabah Fisheries Department Annually Report, 1992, 1993, 1994, 1995, 1996, 1997, 1998, 1999 and 2000)

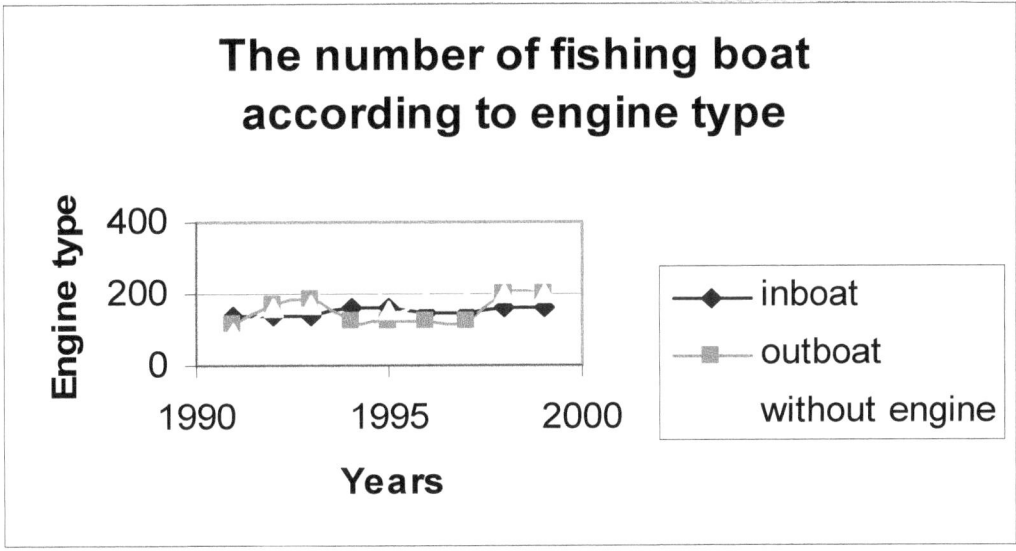

Figure 2.9 The numbers of fishing boat according to the engine type

2.17.10.7 Fish Meal

The production of fishmeal as value added product from underutilized fish, that could be marketed at ruling market price. This is due to the un mature or genuine species underutilized. In year 2000 there are increased by 66.7% from previus year Table 2.7 and Figure 2.10 show the fish meal production.

Table 2.7 The fish meal production

Year	Production (Tones)
1995	8069.04
1996	9644.67
1997	27123
1998	7864.06
1999	2026.64
2000	3378.8

Source: Sabah Fisheries Department Annually Report, (1992, 1993, 1994, 1995, 1996, 1997, 1998, 1999 and 2000)

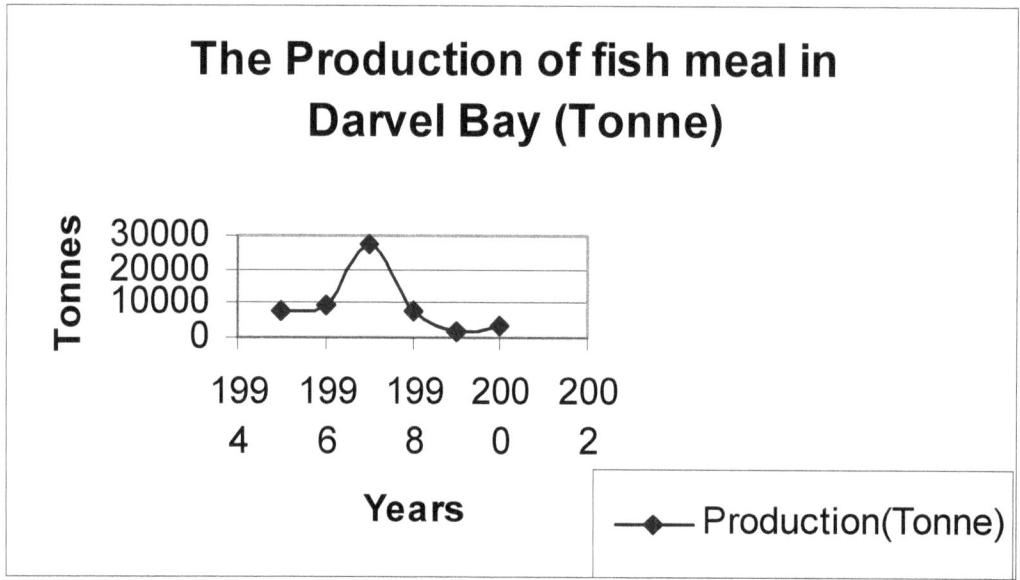

Figure 2.10 The fish meal production.

DESCRIPTION OF THE STUDY AREA

3.1 Malaysia

Malaysia was formed in 1963 through merging of the former British colonies of Malaya and Singapore, including the East Malaysian states of Sabah and Sarawak on the northern coast of Borneo. It is Located in Southeastern Asian peninsula and in northern part of Borneo. Sabah complies one-third of the island of Borneo, which is bordering Indonesia, the South China Sea and south of Vietnam. The area is covered with 329,750 square km water, 1,200 square km land, 328,550 square km with 4,675 km coastline (Peninsular Malaysia 2,068 km, East Malaysia 2,607 km). The continental shelf is 200-meter depth or to the depth of exploitation as specified boundary in the South China Sea. The exclusive economic zone is 200 NM and territorial sea is 12 NM. It has tropical climate with annual southwest (April to October) and northeast monsoon (October to February). The environment included in the international agreements such party as Biodiversity, Climate Change, Desertification, Endangered Species, Hazardous Wastes, Law of the Sea, Marine Life Conservation, Nuclear Test Ban, Ozone Layer Protection, Ship Pollution, Tropical Timber 83, Tropical Timber 94, Wetlands signed, but not ratified: Climate Change-Kyoto Protocol (CIA, 2003)

3.2 Sabah

Sabah is located 1200 km from Kuala Lumpur the capital nation of Malaysia. Separated from West Malaysia by the South China Sea, from the north bounded by Philippines, from the west coast by Brunei Darul Salam and Indonesia in the West. Sabah, a part of Malaysia occupies the top portion of the Island of Borneo. It is also known as the land below the wind, which covers an area of 76,115 km square (29,388 sq. miles) and a coastline of about 1,448 km (900 miles). In the east coast there are South China Sea on the west, the Sulu and Celebes Seas on the east and Sulawesi Sea in the Southern east.

A well known oceanic Island of Sipadan can be found off the north east of Sabah. It was formed million years ago under the sea volcano. The size is only about 4 hectares (approx. 10 acres), lies in cool blue water fringed by sandy beaches, which covered mainly with primary jungle. Exploring Sipadan's underwater, corals grow best in clean, clear water with different shapes and colors from bluish branching colonies to brown plates, orange cup corals and the black coral, highly prized by the jewelry industry. Off the reef face different species of marine habitat can be found. Thousand of large barracudas can be seen circling and moving like an underwater tornado and an impressive diving.

Down to the heart of the state, a low gully known as Danum valley can be reached about 50 minutes drive from the nearby town of Lahad Datu. It is well known for its virgin tropical rainforest, which is mostly populated with flora and fauna of different species. The area is covered by secondary rainforest. It is also a research center from worldwide scientists.

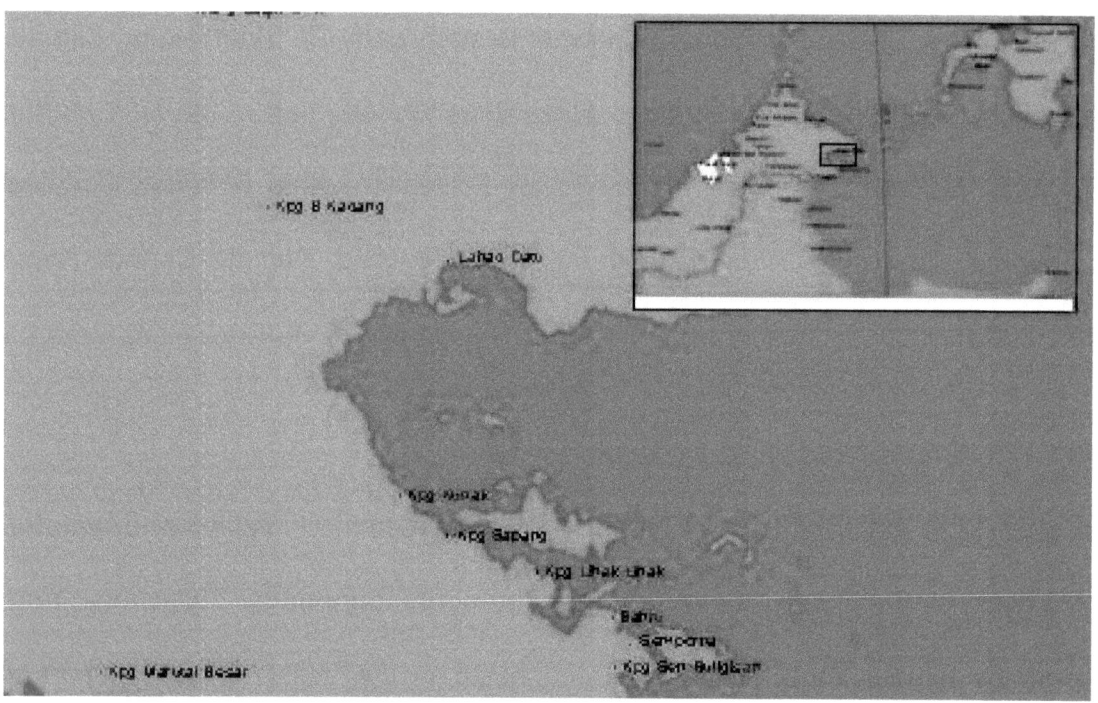

Figure 3.1 The area of Malaysia waters surveyed by stock assessment and population dynamics of *Priacanthus tayenus* (Richardson, 1846) program in Darvel Bay, North Borneo

3.3 Lahad Datu

Lahad Datu is situated in the east coast of Sabah where coastal area along the Darvel Bay face to the Sulawesi Sea in Indonesia and Sulu Sea in the Philippines. The area covered specifically with longitudes 117^0 25'E and 119^0 0'E and latitudes of 4^044'N and 5^0 17'N. There are three main rivers in this district namely; Silabukan, Sepagaya and Kelumpang Rivers.

Lahad Datu is the fourth biggest town in Sabah after Kota Kinabalu, sandakan and Tawau the distance as in Table 3.1. This district is divided into six divisions such as Town area, Silabukan, Dam, Silam, Segama and Tungku including Peninsular Dent. Lahad Datu District is adjecent to Kinabatangan district in the north while Kunak is a town that is injunction with Semporna District in the east.

Table 3.1 The distance of Lahad Datu town to the other neighboring town in Sabah

Town	Distance
Kota Kinabalu	404 Km
Sandakan	154 Km
Kinabatangan	54 Km
Kunak	75 Km
Tawau	149 Km
Semporna	160 Km

Source: District office, Lahad Datu

3.4 Darvel Bay

The Darvel Bay is a major fishing ground for aquaculture, rich in marine resources, impact on marine environmental, Darvel Bay is also well known for its potential pirates to operate and treat the fishermen in the coastal areas.

Darvel Bay, the biggest bay in Sabah is situated in the east coast of North Borneo (Figure 3.2). It has a long coastline lying from Tungku to Lahad Datu and further down to Kunak district which dominants by small islands where fishing activities being done in their adjacent waters.

The study was conducted in Darvel Bay due to largest number of fishermen involve and fishing activities such as purse seining, trawler being practiced in this area. The coastline of Darvel Bay cover about 113 miles extends from north-east Tungku to east of Kelumpang River in Kunak. Most of the population is concentrated along the coastline including Lahad Datu and Kunak town.

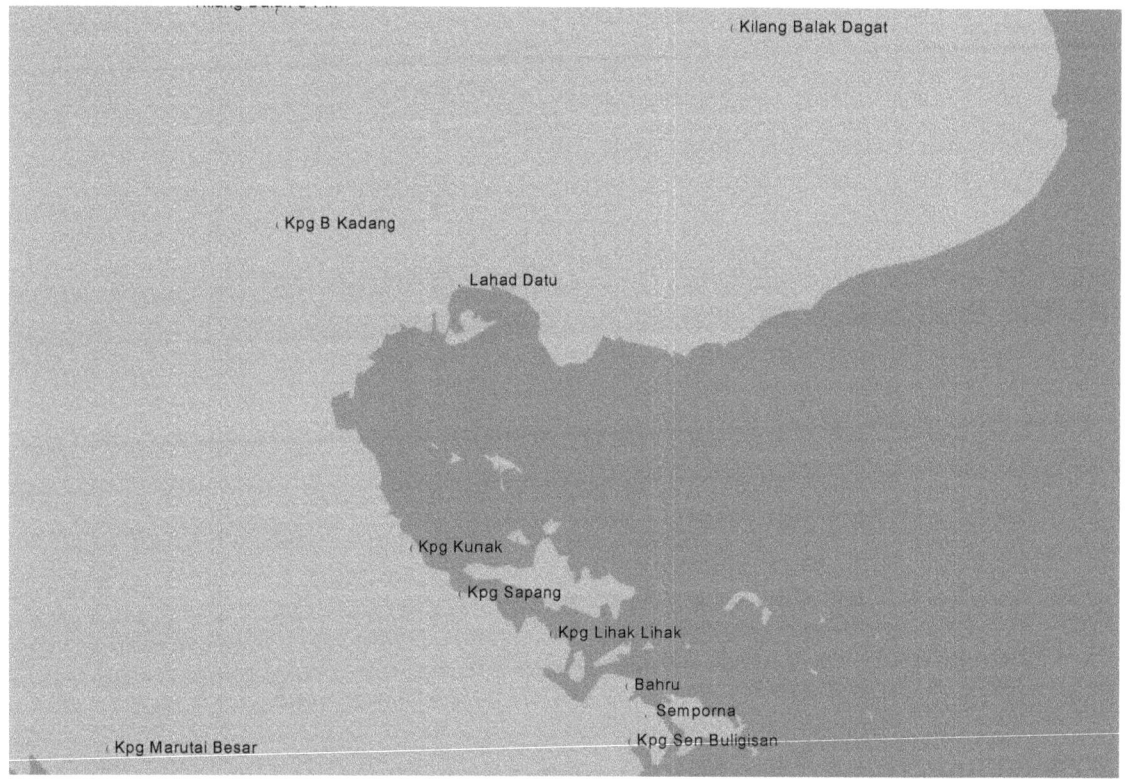

Figure 3.2 The area of Darvel Bay and other neighboring town in Sabah, Malaysia

The two landing places such as of jetty nearby fish market in Darvel Bay are Lahad Datu in north and Kunak in the southeast. These two districts were chosen as the main areas for sampling activities.

3.5 Topography

The land area of Lahad Datu district is about 687.372 ha or 6500 Kilometer square and about 320.817 hectares are suitable for agriculture. The Danum valley international research centre in Segama is popular for its fauna and flora habitat. The Tabin wild life research centre is located in Begahak Range 760 meters wide is famous for its wild life animal species.

3.6 Environmental Parameters

The climate of Darvel Bay was based on the meteorological data from District of Lahad Datu. The environmental parameter was obtained from the Darvel Bay during the study period from November 2001 to October 2003.

3.6.1 Depth

By using ecosounder the depth ranging from 5 meters to 40 meters and map produced by Marine Department of Sabah in title Sabah – East Coast DARVEL BAY No. 1680 dated on 2002 published by Taunton, United Kingdom as a guide while at sea.

3.6.2 Potenz Hydrogen (pH)

The variation of Potenz hydrogen (pH) is noted between 8.67-8.86 near the shore and between 8.65-8.83 from the selected stations sited in open sea.

3.6.3 Monsoon Seasons

Throughout the year, this area is effected by wind from the northeast and southeast or northeast and southeast monsoons. The southwest monsoon (April to October) and northeast (October to February) monsoons. Darvel Bay has a typical monsoon and commonly found during both the northeast and southeast monsoon seasons. This phenomenon is caused by differences in temperature between continental and oceanic areas. Annually two monsoons seasons, the Northeast monsoon and Southeast monsoons.

3.6.4 Rain

The rainy season lasts from November to March, reaching its maximum in November and December. The dry season occurs from May to October, with a minimum in June. The average rainfall is 1899 mm during the northeast monsoon, while 1015 mm during the southeast monsoon. The average annual rate is 1457 mm.

3.6.5 Sea Surface Salinity

The variation of sea surface salinity is between 32-33.2 ppt in coastal area and high seas. During the peak of northeast and southeast monsoons the average sea surface salinity is 33.5 ppt.

3.6.6 Temperature

The climate is relative hot and humidity of 80.5% and the temperature range from 26 to 31O C. The yearly sea surface temperature is low between 28.8 300C.

3.6.7 Winds

The winds blow from south-west and are more regular. The wind blows from the Northeast and southeast direction has maximum 50 km per hour. This condition may prevent all fishing activity during the Northeast monsoon that occurs during November and March. It is characterized as very windy periods with frequent rainfalls lasting for days. The south-east monsoon occurs from May to September while the wind blows from the north-east in November to March.

3.6.8 Waves

The wave travel on the surface of the sea, between the highest point of wave or crest. The lowest point between the two crests is called the trough. According to Malaysian

Department of Climatology, Petaling Jaya the maximum waves reach 3 meters height occur during rough seas. The coastal water of Sabah is relatively moderate in north-east monsoon. The swell may reach two until three meters height during north-east monsoon in August and relatively calm in February. Most of the area having less than one meter wave heights when influenced by the north-east monsoon

.

3.6.9 Tidal

The movements toward and away from the shore are the effects of tides. Tides are caused by the gravitational attraction between the moon and the earth. The influence of tide on the current patterns does not show any strong oscillation resulting in reverse currents. The water is flowing consistently in northwesterly direction with periodic deviations to both sides following the tide cycle. According to Sabah Marine Department, high tides were recorded 2.4 meter at 1807 hour and 1849 hour and as low as 0.0 meter at 12.02 hour on 24 and 25 November, 2003 (Appendix 9)

3.6.10 Hydrology

The main factor influencing oceanographic conditions of the Darvel Bay is Sulu Sea current. It has a tendency to deflect from the shore, making it possible to move out from the bay to the high seas from time to time. The changing patterns of monsoon winds influence the current patterns in this area. The velocities of Darvel Bay current are changeable, usually flowing from 20 to 40 cm/s in the surface current only.

3.6.11 Current

The waves that are raised by the winds carry with them a great deal of energy while the wind produces a continuous slow movement of the water on the water surface. The current likely influenced by the changing patterns of monsoon winds, off northwest of Darvel Bay. Eastern current flows from November to March and reverses to North West current from may to October. The month of transition is between April to May and September to October. The current speeds are generally higher during low tides comparing to high tides or slack period. Such behaviors are largely attributed to the configuration of surrounding landmasses and the sea bottom.

3.7 Economics

The Lahad Datu economics is mostly based on agricultural products such as oil palm, cocoa, rubber and short term cultivation. Other sources are fisheries, veterinary and timber industries.

3.7.1 Agriculture

Lahad Datu, one of the most fertile soils in Sabah has a relative flat area suitable for cultivation especially in Segama valley. The Segama highland and along its river has an alluvial type of land which has come to be the most fertile land in Sabah. The formed of volcanic soil are suitable for oil palm, cocoa industry that contributes to the economic of the district and state as well.

3.7.2 Veterinary

Veterinary industry in this district is familiar locally especially indigenous community. A cattle farming was first introduced in Segama area during British colonial. Other common small-scale animal industry such as goat, pig, poultry and other birds has been practiced in the area. Nowadays potential integrated farming is popular among large oil palm plantation.

The veterinary industries have a bright future in this district due to the existence of large oil farm plantation. The palm oil kernel cake (PKC) is largely available from oil palm processing as value added and can be utilized as animal feed. The veterinary populations in Lahad Datu District are presented in Table 3.2.

Table 3.2: Veterinary population in Lahad Datu District in 1996-2000 (heads)

Type of farm	1996	1997	2998	1999	2000
Buffaloes	1652	2153	2166	2509	3556
Cow	1652	1568	1632	1800	1988
Goats	3073	2601	2965	2655	2591
Pigs	3100	2219	3100	3213	3504
Poultry	49600	11455	11000	5000	4500

Source: Annually Report Sabah Veterinary Department, 2001

3.7.3 Fisheries

The marine resources off the east coast of Sabah has been exploited by small scale fisheries since long time by using large variety of fishing gears. From surveys that had been carried out especially on demersal resources study, it was found a least 72 species of fish can be collected from this the area.

The collected data of total marine fish landings in Darvel Bay from Kunak and Lahad Datu districts in 2003 is tabulated in Table 2.3 had contribuated 14.8% annually to marine fish landing in Sabah.

MATERIALS AND METHODS

4.1 Duration and Location of Study Area

The study was conduced from November 2001 to October 2003 in Darvel Bay, Sabah, Malaysia. Data analyses were done in the College University of Science and Technology Malaysia, (KUSTEM), Kuala Terangganu, Malaysia.

4.2 Description of the Study Area

The area of study is included in the three main districts of Lahad Datu, Kunak and northern part of Semporna. The locality, specifically in position within the boundaries of 118° 10´ to 118° 40´ East longitude and 5° 0´ to 4° 40´ North latitude. A Global Position System (GPS) (Plate 4.1) device was used to locate the position of the stations. There were 128 stations designed for trawler as tabulated in Table 4.1. Six islands were involved such as Pulau Tebawan, Pulau Batik Kelambu, Pulau Timbun Mata, Pulau Bohayan, Pulau Maganting Pulau Sakar. The locations of the study are shown in Figure 4.1.

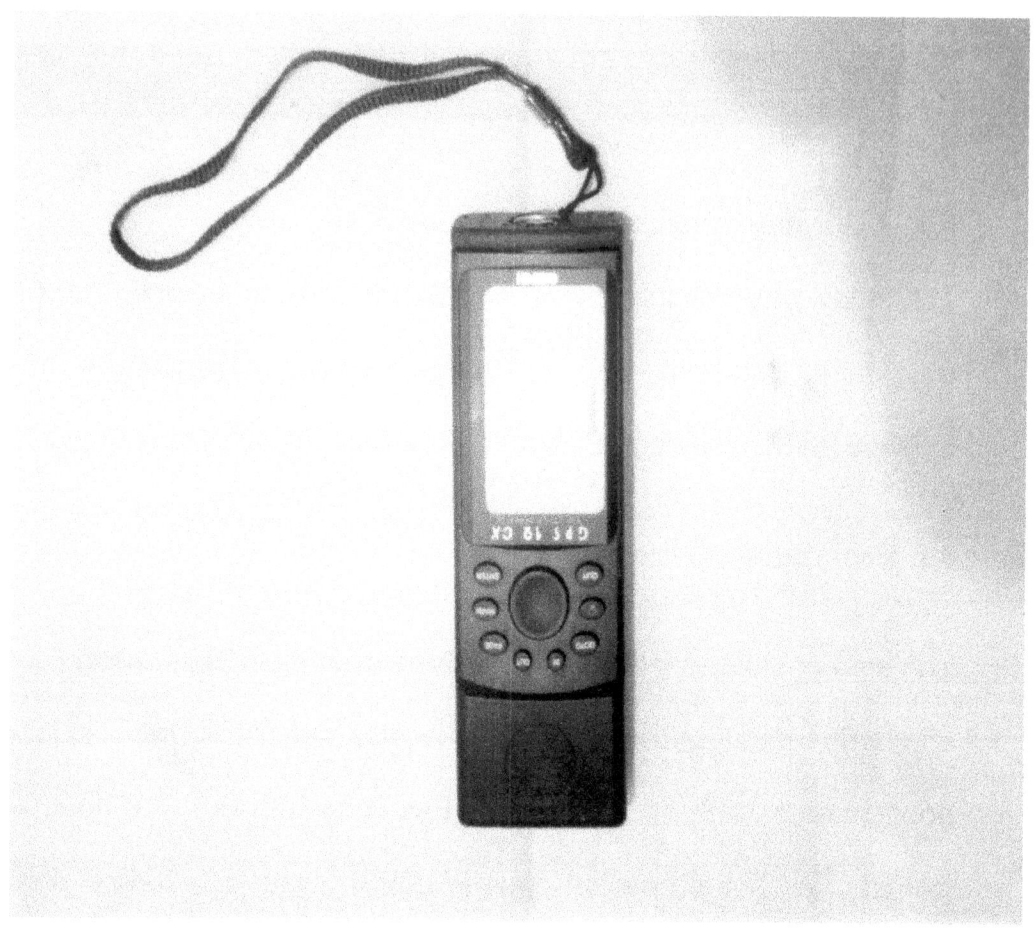

Plate 4.1 Global positioning system (GPS) to locate the position of stations

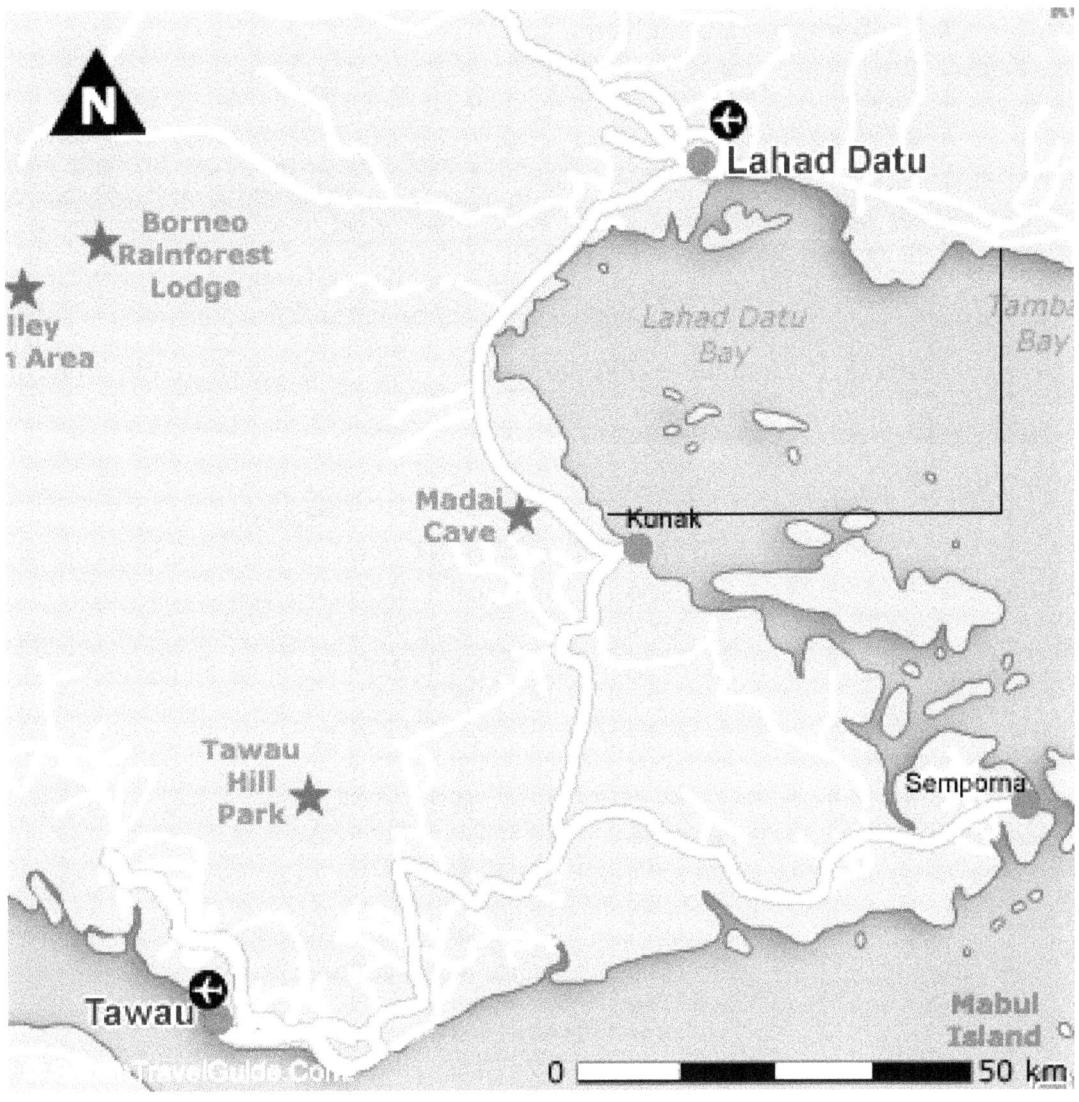

Source: Sabah Marine Department, 2002

Figure 4.1 Chart of Malaysian waters showing the area sampled for *Pariacanthus tayenus* in Darvel Bay, Sabah, Malaysia in the Sulu-Sulawesi Marine Ecology (SSME) region

4.3 Samples Collection

The data were collected based on fish species of Priacanthus tayenus . were caught from November 2002 to October 2003, in Darvel Bay as in Table 4.1 and Figure 4.1. A commercial vessel (Plate 4.2) with a capacity of 190 Horse Power (HP) and speed of 3 Knots was used. It was equipped with a trawler net of cod end mesh size 38 mm. (Plate 4.2), the area are shown in Figure 4.2.

Table 4.1 Longitude and latitude of the 128 location stations of sample collected

No	Position
1	Long 118° 24' E, Lat 4° 50' N
2	Long 118° 14' E, Lat 4° 53' N
3	Long 118° 25' E, Lat 4° 58' N
4	Long 118° 15' E, Lat 4° 53' N
5	Long 118° 26' E, Lat 4° 57' N
6	Long 118° 16' E, Lat 4° 52' N
7	Long 118° 27' E, Lat 4° 57' N
8	Long 118° 15' E, Lat 4° 52' N
9	Long 118° 25' E, Lat 4° 57' N
10	Long 118° 16' E, Lat 4° 52' N
11	Long 118° 26' E, Lat 4° 56' N
12	Long 118° 14' E, Lat 4° 51' N
13	Long 118° 12' E, Lat 4° 55' N
14	Long 118° 13' E, Lat 4° 47' N
15	Long 118° 14' E, Lat 4° 48' N
16	Long 118° 15' E, Lat 4° 54' N
17	Long 118° 14' E, Lat 4° 47' N
18	Long 118° 13' E, Lat 4° 53' N
19	Long 118° 16' E, Lat 4° 46' N
20	Long 118° 12' E, Lat 4° 54' N
21	Long 118° 18' E, Lat 4° 48' N
22	Long 118° 11' E, Lat 4° 53' N
23	Long 118° 24' E, Lat 4° 55' N
24	Long 118° 33' E, Lat 4° 57' N
25	Long 118° 26' E, Lat 4° 59' N
26	Long 118° 32' E, Lat 4° 52' N
27	Long 118° 23' E, Lat 4° 56' N
28	Long 118° 35' E, Lat 4° 53' N
29	Long 118° 53' E, Lat 4° 23' N
30	Long 118° 34' E, Lat 4° 52' N
31	Long 118° 23' E, Lat 4° 53' N
32	Long 118° 32' E, Lat 4° 52' N
33	Long 118° 23' E, Lat 4° 58' N
34	Long 118° 30' E, Lat 4° 53' N
35	Long 118° 23' E, Lat 4° 57' N
36	Long 118° 30' E, Lat 4° 52' N
37	Long 118° 23' E, Lat 4° 56' N
38	Long 118° 30' E, Lat 4° 54' N

39	Long 118° 23' E, Lat 4° 55' N
40	Long 118° 30' E, Lat 4° 55' N
41	Long 118° 23' E, Lat 4° 54' N
42	Long 118° 31' E, Lat 4° 55' N
43	Long 118° 22' E, Lat 4° 57' N
44	Long 118° 14' E, Lat 4° 51' N
45	Long 118° 22' E, Lat 4° 56' N
46	Long 118° 13' E, Lat 4° 50' N
47	Long 118° 22' E, Lat 4° 56' N
48	Long 118° 15' E, Lat 4° 54' N
49	Long 118° 22' E, Lat 4° 52' N
50	Long 118° 13' E, Lat 4° 53' N
51	Long 118° 22' E, Lat 4° 53' N
52	Long 118° 12' E, Lat 4° 54' N
53	Long 118° 21' E, Lat 4° 57' N
54	Long 118° 14' E, Lat 4° 55' N
55	Long 118° 21' E, Lat 4° 56' N
56	Long 118° 14' E, Lat 4° 57' N
57	Long 118° 21' E, Lat 4° 55' N
58	Long 118° 14' E, Lat 4° 52' N
59	Long 118° 21' E, Lat 4° 54' N
60	Long 118° 13' E, Lat 4° 53' N
61	Long 118° 21' E, Lat 4° 53' N
62	Long 118° 14' E, Lat 4° 53' N
63	Long 118° 20' E, Lat 4° 53' N
64	Long 118° 14' E, Lat 4° 52' N
65	Long 118° 23' E, Lat 4° 57' N
66	Long 118° 14' E, Lat 4° 53' N
67	Long 118° 23' E, Lat 4° 58' N
68	Long 118° 15' E, Lat 4° 53' N
69	Long 118° 23' E, Lat 4° 56' N
70	Long 118° 16' E, Lat 4° 52' N
71	Long 118° 23' E, Lat 4° 55' N
72	Long 118° 16' E, Lat 4° 53' N
73	Long 118° 23' E, Lat 4° 54' N
74	Long 118° 16' E, Lat 4° 52' N
75	Long 118° 22' E, Lat 4° 56' N
76	Long 118° 14' E, Lat 4° 55' N
77	Long 118° 12' E, Lat 4° 55' N
78	Long 118° 13' E, Lat 4° 47' N
79	Long 118° 14' E, Lat 4° 48' N
80	Long 118° 12' E, Lat 4° 54' N
81	Long 118° 14' E, Lat 4° 47' N
82	Long 118° 13' E, Lat 4° 53' N
83	Long 118° 16' E, Lat 4° 46' N
84	Long 118° 12' E, Lat 4° 53' N
85	Long 118° 18' E, Lat 4° 48' N
86	Long 118° 11' E, Lat 4° 53' N
87	Long 118° 24' E, Lat 4° 55' N
88	Long 118° 13' E, Lat 4° 57' N
89	Long 118° 26' E, Lat 4° 59' N
90	Long 118° 13' E, Lat 4° 52' N
91	Long 118° 23' E, Lat 4° 56' N

92	Long 118° 14' E, Lat 4° 53' N
93	Long 118° 53' E, Lat 4° 23' N
94	Long 118° 14' E, Lat 4° 52' N
95	Long 118° 23' E, Lat 4° 53' N
96	Long 118° 13' E, Lat 4° 52' N
97	Long 118° 25' E, Lat 4° 56' N
98	Long 118° 14' E, Lat 4° 55' N
99	Long 118° 24' E, Lat 4° 54' N
100	Long 118° 15' E, Lat 4° 56' N
101	Long 118° 27' E, Lat 4° 54' N
102	Long 118° 16' E, Lat 4° 56' N
103	Long 118° 26' E, Lat 4° 56' N
104	Long 118° 15' E, Lat 4° 57' N
105	Long 118° 26' E, Lat 4° 56 N
106	Long 118° 16' E, Lat 4° 51' N
107	Long 118° 25' E, Lat 4° 56' N
108	Long 118° 12' E, Lat 4° 51' N
109	Long 118° 14' E, Lat 4° 50' N
110	Long 118° 14' E, Lat 4° 43' N
111	Long 118° 13' E, Lat 4° 47' N
112	Long 118° 14' E, Lat 4° 47' N
113	Long 118° 15' E, Lat 4° 54' N
114	Long 118° 16' E, Lat 4° 46' N
115	Long 118° 13' E, Lat 4° 53' N
116	Long 118° 18' E, Lat 4° 48' N
117	Long 118° 12' E, Lat 4° 54' N
118	Long 118° 24' E, Lat 4° 55' N
119	Long 118° 11' E, Lat 4° 53' N
120	Long 118° 26' E, Lat 4° 59' N
121	Long 118° 33' E, Lat 4° 57' N
122	Long 118° 23' E, Lat 4° 56' N
123	Long 118° 32' E, Lat 4° 52' N
124	Long 118° 53' E, Lat 4° 55' N
125	Long 118° 35' E, Lat 4° 21' N
126	Long 118° 32' E, Lat 4° 51' N
127	Long 118° 34' E, Lat 4° 54' N
128	Long 118° 31' E, Lat 4° 53' N

Plate 4.2   The commercial vessel used in this study

Plate 4.3   The trawler net with cod end mesh size 38 mm

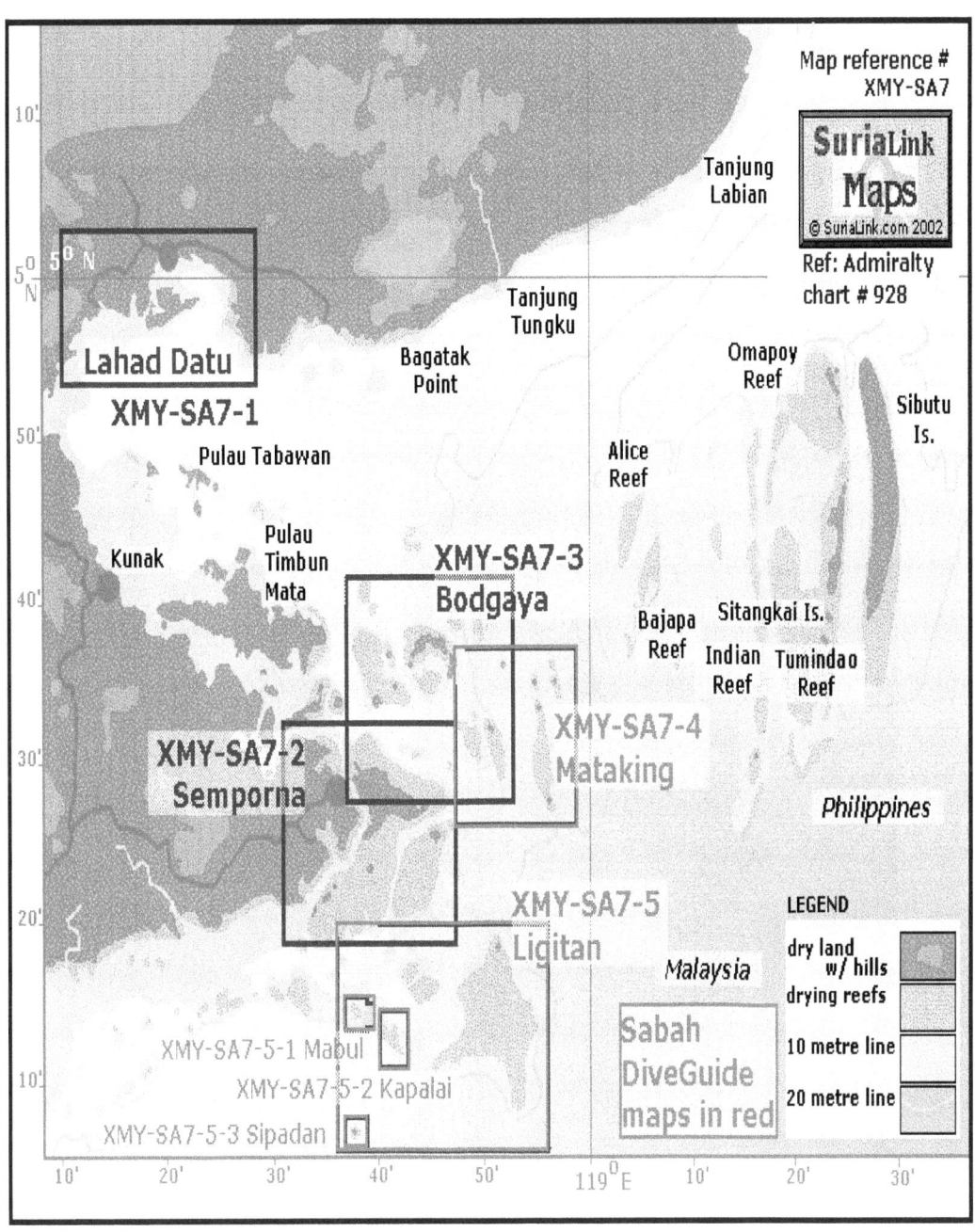

Figure 4.2 The survey area using Swept Area Method in Darvel Bay

4.4 Catch

Fish samples were caught using trawler as in Plate 4.3, which ranged within 4 hours for each trawl. The samples were measured for their length frequency as in Plate 4.4 and were then kept in a container.

Plate 4.4 The *Priacanthus tayenus* were caught by trawler

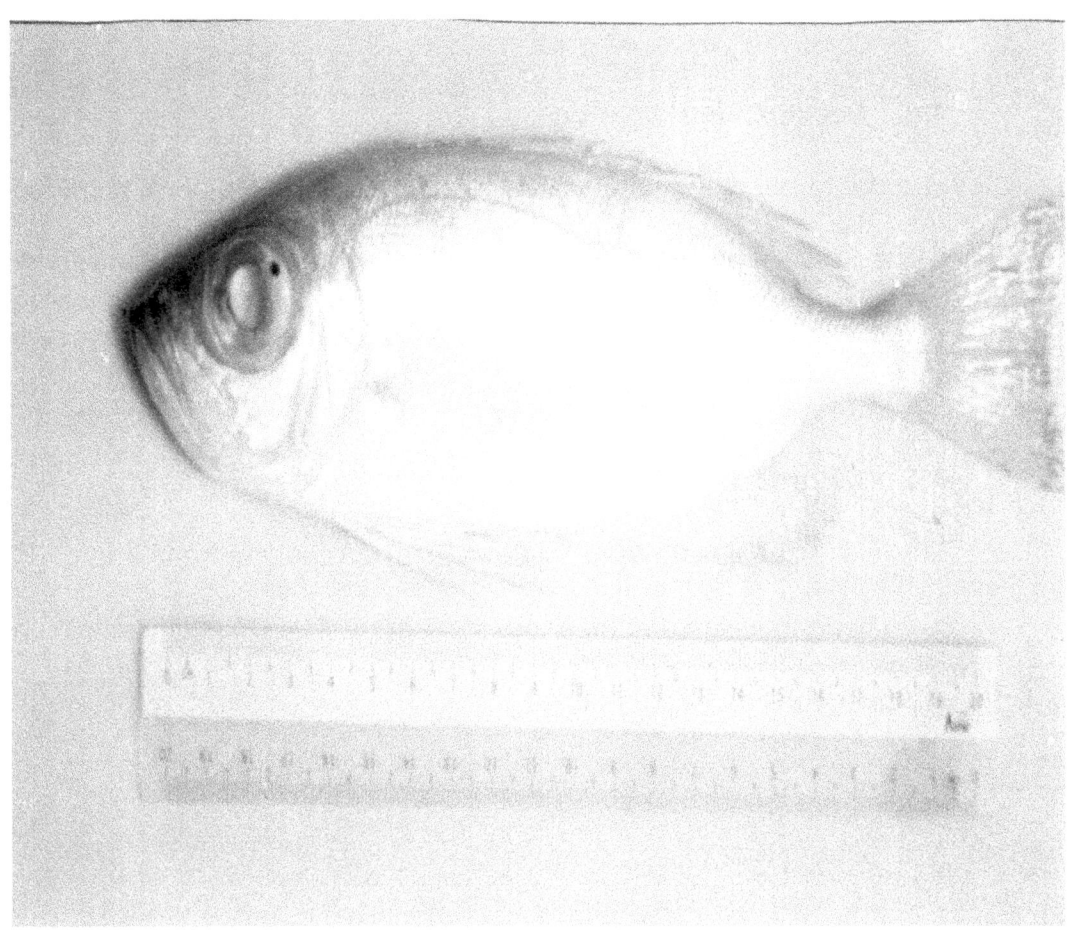

Plate 4.5   Measurement of *Priacanthus tayenus* sample

4.4.1 Store

The unmeasured or unweighed samples were packed in polythene bags before being stored in a temperature of 4^0 C until measurements and weighing were completed.

The same procedure was applied to all different types of study. It was noted that the study covered the demersal fish from the catch-using trawler.

4.4.2 Species Identification

In order to get a confirmation of the species, samples of *Priacanthus tayenus* (Plate 4.3) were sent to the Ichthyology laboratory in the University College Science and Technology Malaysian (Kustem) for identification.

4.5 Weights and Measure Frequency Length

The weighting instrument (Plate 4.6) and measurement board (Plate 4.7) were used to weigh and measure the length of 24,653 fish samples as shown in Appendix 9 and Appendix 10.

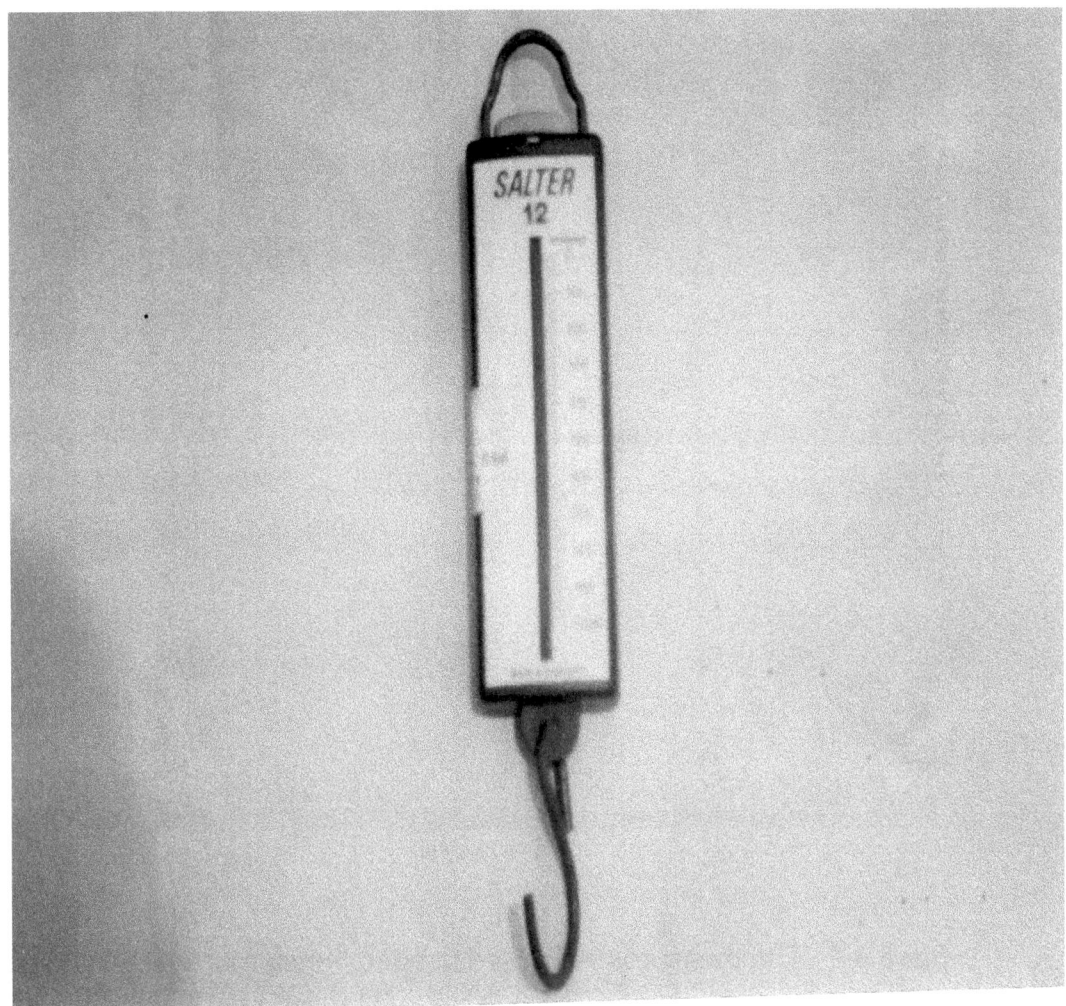

Plate 4.6 Weighting instrument "Salter" to weight the samples

Plate 4.7 The measurement board used to measure frequency length

4.6 General Methodology

The general method was divided into four sections and conducted in Darvel Bay. The biomass dynamics data was provided by Sabah Fisheries Department and other agencies engaged in fishing activities. This was done for planning purpose in the study area.

This study was divided evenly into four blocks according to priority assessment in the area. Further, was statistical analysis of data done using a microcomputer to calculate each method according to the experimental need, results and evaluation. There were two related parts of study in the Darvel Bay as follow;

4.6.1 Survey on the Demersal Marine Fishes

The resources of the area was surveyed to assess the status of Stock. Commercial vessel (Plate 4.2) was used to do the following as;

a. Survey effective area

b. Identification and species percentage

c. Catch ratio of marketable with underutilized and,

d. Survey of demersal fish resources by using Swept Area Method

In this survey the species composition and species caught by trawler were studied. The commercial vessel was used. The length, beam, gross tones (GRT), engine type and measurement of code end mesh size were recorded in each trial.

The lengths of the fresh fish samples were measured and weighed. Both the standard length (cm) and weight (kg) were set in one decimal point. The flow chart experimental procedures were as in Figure 4.3;

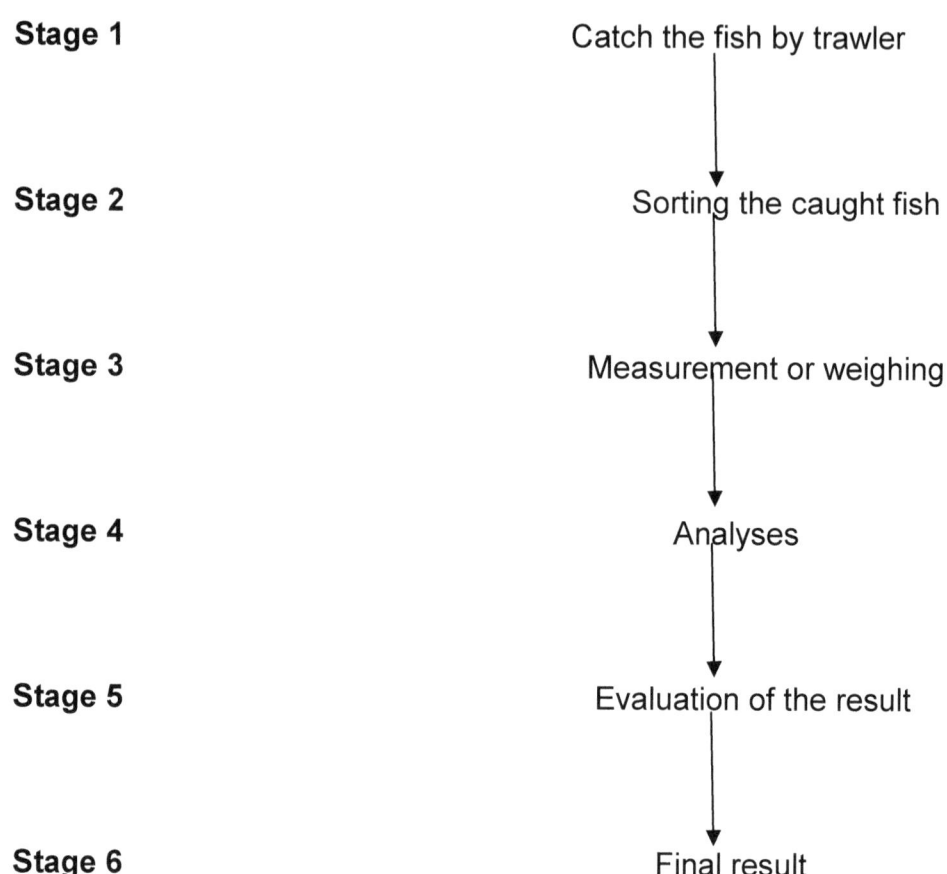

Figure 4.3 The flow chart of the experimental procedures on the resource survey

Plate 4.8 The *Priacanthus tayenus* used in this study

4.6.2 Hamid Awong Fisheries Model (HAFM) for *Priacanthus tayenus* for 50 Years

The colleted data and results from previous study were used to provide the biomass model of *Priacanthus tayenus* for 50 years.

This model takes into account the recruitment, growth, mortality caused by fishing activities and natural mortality. Weight increased though growth and number decreased through mortality either fishing mortality or natural mortality. Assuming an equal fishing effort, the model could project the catch and the profit for 50 years period. The model is designed for policy makers to review their policies.

Determination of the resource status by using parameters from the previous study can forecast future management and suggestion to maintain the sustainable stock.

This biomass model for 50 years could show the parental biomass, which falls to critical level when recruitment could not recover the stock due to high mortality by changing the mortality

The Status Model of *Priacanthus tayenus* according to the flow chart of the experimental procedures are presented in Figure 4.6.

HAMID AWONG FISHERIES MODEL (HAFM)

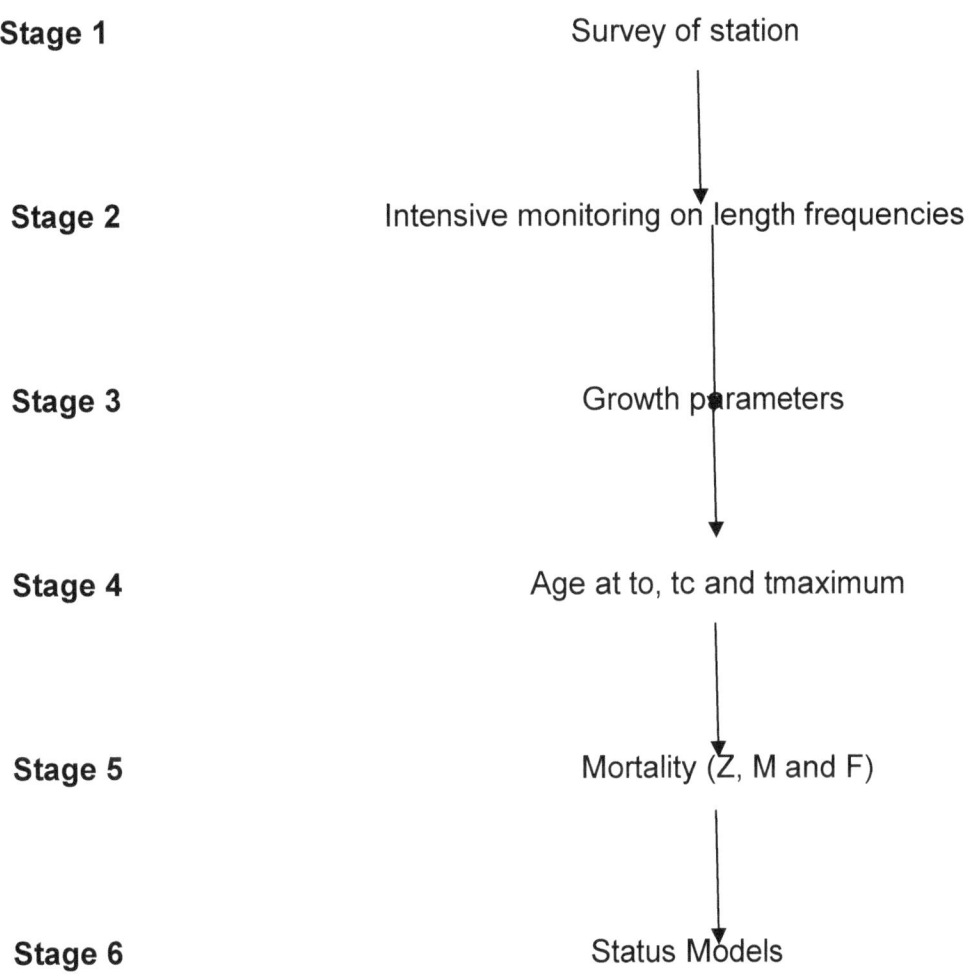

Figure 4.4 Flow chart of experimental procedures on Status Model of *Priacanthus tayenus* for the next 50 years

4.7 Data Analyses

The data were analysed and processed with relevant formula, which was accepted internationally and used by previous scientists in related fields.

4.7.1 Data Evaluation

Data evaluation was done at the departmental seminar

4.7.2 Progress Seminar

Progress seminar was held in presence of the public of angler community or fisherman.

SURVEY ON THE DEMERSAL MARINE FISHES BY TRAWLER

5.1 INTRODUCTION

Darvel Bay is among the rich fishing ground situated in East Coast of Sabah. It's covered the Lahad Datu and Kunak District with an area of approximately 834.55 Kilometer Square Lies within 5 fathom (9.15 Meters) line and above to 12 nautical miles (19.2 Km) along the coastal state and is considered as trawlerable ground. The fishing gear can be divided into commercial and artisanal. The commercial fishing gear including purse seining, trawler and other active gear whiles the arisanal fishing gear as selambau, kelong, pole and line, gill net trap and bagang.

Various type of gear can be found from simple and traditional to commercial fishing gear such as trawler. The fishing area for trawler lies within a depth of 5 fathoms (9.15 Meter) line and above. The others are traditional gear mostly on the bed sea mud of trawler able.

The trawler with 7 vessels was introduced into this State in 1971 (Fisheries Dept. Annually Report, 1972). It was the beginning of new era in fisheries sectors in the state. In the beginning the impact of high catch and operation to the environment and fishermen were small and the owner gained high profit. The trawler as a commercial fishing gear was introduced in 1975 in Darvel Bay where prawn as a target. Some modification was made on this fishing gear for catching demersal fish in the coastal area.

As a largest bay in Sabah, Darvel Bay is important to the local as source of food as well as to get cheap protein. Darvel Bay is known for fisheries activities and as a major fishing ground where most of the community in the coastal area are full time or part time fishermen.

The commercial fishing gear had increased demersal fish landing in this area. Figures were obtained from statistical annually data which was collected in Lahad Datu and Kunak district. In 1990 the marine fish landing was 13761 tones and this amount increase to 53051.36 tones after 10 years. The. Increasing of landing was due to the increasing of fishermen and fishing gears. The catch trend increases by quantity every year because of increase in effort fishing gear. The fishery manager can be able to use the catch trend as a further rational planning for expansion in future. This fishing gear can be expanded into industry with sustainable manner to the environmental.

In 1991 the full time fishermen were 700 and had increased to 1470 in 2000. The fishing gears in 1991 were 282 and 795 in 1999 (Fisheries Department Annually Report, 2000). A number of 31-trawler vessels operations in this area as included in the figure. The catches by commercials fishing gear were 23,259 tones while traditional fishing gear 38,028.84 tones. Data that available from commercial catch as secondary data and the survey data or primer data can be used for this purpose.

This study is designed to improve understanding of the status of stock resources in this area. The catch by most of the fishing gear especially trawler is considered low. Most of the catch were non-targeted fish or immature although included in commercial species. However due to under market size or unmarketable and valueless in view of quality the catch is considered underutilized. Change in the catch composition by trawler can be used as an indicator that over fishing has occurred in this industry within this area.

The objective of the survey is to assess the potential of demersal fish resources and to identify the areas suitable for trawling in order to implement development plans of commercial trawling in Darvel Bay. The data provided from the survey as a basis for assessing the magnitude of the resources and its resilience to fishing pressure.

5.2   MATERIALS AND METHODS

The objective of the demersal research was to survey the resources within the area of Darvel Bay, to assess the status of the stock and the ratio of commercial and underutilized fish because of under market size and unmarketable. These assessments in particular provide basis for rational management in this area as well as in Sabah generally in future.

A 20 GRT trawler was used in the trial to collect fish samples. Extensive experiments had been held in 2000 to 2002. The tow duration was 4 hours by an engine 1100 rpm at 3 knots speed. The samples were taken on board with registration number LD. 1192/F and the code end mesh sizes were 38 mm. A total of 128 hauls were conducted within the area ranging between 5-60 fathoms (9-108 Meters) depth. Data of total weight catch were recorded for each haul. The map of Darvel Bay No. 1680 that produced by Superintendent of Rear Admiral, United Kingdom was used during the survey as in Figure 4.1. It was used to calculate the approximate area of the bay as stated in Table 5.1.

The trawlers activities are summarized in Table 5.3 One hundred and twenty-eight separate locations were involved in trawling activities using the trawler boat. An alternative fishing ground was allocated during bad weather in the early survey.

5.2.1 Research Location

Darvel Bay situated in Sabah is located along the east coast covered with 5-60 fathoms (9-105 meters) depths, generally regarded as highly productive area. The area is approximately 834.55 kilometer square lying within the boundaries of Longitude 118° 10´ East to Longitude 118° 40´ East and Latitude 5° 0´ to 4° 40´ North. It has a long coastline and shallow fresh waters flowing from Silabukan, Sepagaya, Tingkayu River that provide a large amount of habitat of marine resources.

5.2.2 Methods

5.2.2.1 The Effective Survey Area

The map produced by Marine Department was used to identify the target area for the purpose of surveying the resources. The target area was divided into 17 blocks and each block had an effective area for trawling.

5.2.2.2 Identification and Species Percentage

The Collected samples were identified according to their family or species. The percentage for each family or species was also noted and calculated. Sorting was carried out as soon as the catch landed on deck.

5.2.2.3 The Catch Ratio of Marketable and Underutilized

The catch shortages based on market size and marketable species were calculated on percentage of marketable and non-marketable or underutilized.

The fish that had commercial value were sorted out according to their species.

5.2.2.4 The Demersal Fish Resources

5.2.2.4.1 Swept Area Method

Global Position Sensory (GPS) was used to identify the location of 128 stations that involved in trawling. The trawler net was used to estimate the mean catch in the fishing area at all station. The size of stock or total stock weight or biomass was the mean catch per area being swept by trawl (a) multiplied by the stock area (A). A towed trawl net in fact the fish sample in an area which is equivalent to a long rectangular sampling unit with an area, (Spare and Siebseh, 1992: King, 1996) estimated as:

$a = W * TV * D$

$W = HL/N$ ml (1852 metres))*TV*Duration

Where;

W = the effective width of the trawl

HL = head line (Meter)

N ml = Nautical Mile, where's 1 N mile = 1852 Nautical meters

TV = the velocity (M/hr)

D = duration of the tow (hour)

With assumption the doors and sweeps are effectively herding fish into the path of the net. Table 5.1 shown the information data for Swept Area Method. The weight or stock biomass of fish in the area is obtained by multiplying the biomass of the fish in the path of the trawl by the ratio of the stock area to the trawl area:

$$B = Cw/v * (A/a)$$

Where;

    B = Biomass or stock size

    Cw = Mean catch per tow

    v = Vulnerability (Fish above 5 cm Vulnerability with mesh size 38 mm)

    A = Total area occupied by the stock

    a = The area covered by the standard trawl

Table 5.1 Information Data for Swept Area Method

1 nautical M	1852	Meters
Towed distance	24456.8	Meters
Fishing area	834.55	Km.sq
Headline	6.45	Meters
Towed velocity	6.08	Km/hrs
Time	4	Hrs
Mean catch	162.6	Kg

CHAPTER - 5

5.3 RESULTS AND DISCUSSIONS

5.3.1 The Survey Effective Area

The survey area was divided according to blocks, with consideration of trawl able fishing area 834.55 kilometers square

5.3.2 Species Identification and Weight of each Species

The collected sample caught using trawler boat was arrange by using International Standard Statistical Classification of Aquatic animals and Plants (ISSCAAP) format and identified according to their species. There were 71 species being classified according to their International Standard Statistical Classification of Aquatic animals and Plants (ISSCAAP), local and English name, Scientific name and the weight of each group. The dominant species were Kerisi or Thereadfin bream (*Nemipterus spp*) 1001.645 kilograms, Duri/Pulutan/Utek or Marine catfish (*Tachsurus spp./Arius spp,Osteogenius spp*) 1001.645 kilograms or 5%, Batu or Parrot/Wrass (*Labridae scaridae*) 801.316 kilogram, Kerapu or Grouper (*Epinephelus spp*) 801.316 kilograms, Beliak mata/lolong bara or Thereadfin big eye (*Prianchantus tayenus*) 801.316 kilograms or 4%, Gelama/Tengkerong or Jewfish (*Sciaene spp./Johnius spp*) 600.987 kilo grams, Kerisi bali or Sharptoothed bass (*Pristipormoides typus*) 600.987kilograms, Kikek or Ponyfish (Slipmounth) or *Leiognathus spp*. 600.987 kilograms, Chelek mata or blotched grunt (*pomadasys spp*.) and Selangat or Bony bream *(Anodontostoma chacunda)* 3% Belanak Mullet *Liza spp./yalamugi spp*, Tenggiri and *penaeid* 3% respectively as in Table 5.2.

Table 5.2 Species Composition According International Standard Statistical Classification of Aquatic animals and Plants (ISSCAAP) format with 128 Hauls by trawler

ISSCAAP Code	Division Group of Species	Local Name	English Name	Scientific Name	Weight (Kg)
24	Shad, Milk-	Kebasi/	Chacunda	Anodontosoma	180

Fishes, Etc.	Selangat	shad		
	Puput	Shad	Pellona spp.	371
	Terubok	Longtail shad	Shad hilsa macrura	140
25 Miscellaneous	Siakap	Giant sea perch (Barramudi)	Lates calcarifer	80
31 Flounders, Halibut, Soles, Etc	Sebelah	Flatfish	Pseudorhombus spp.	391
	ikan buaya	Flathead	sauride spp	158
	ik.terbang	Flying fish	exocoetidae	20
	Tempaku	Mud skipper	gobioidae	40
	Anjang anjang	Threadfin bream	Pterocaesio spp	30
33 Redfishes, Basses, Congers, Etc.	Batu	Parrot/Wrass	Labridae scaridae	771
	Delah/sulit	Fussiller	Caesioerythrogaster/ C.chrysona	353
	Duri/Pulutan/Utek	Marine catfish	Tachsurus spp./Arius spp,Osteogenius spp	962
	Gelama/Tengkerong	Jewfish	Sciaene spp./Johnius spp. Osteogenius spp.	561
	Gerut-gerut	Grunter	Promadasys spp.	341
	Jenahak	Mangrove snapper	Lutianus johni	160
	Kerapu	Grouper	Epinephelus spp.	781
	berengan/hantu	Grouper	Epinephelus spp.	60
	maming/licin		Epinephelus spp.	26
	kerapu tikus	Grouper	Epinephelus spp.	50
	Kerisi	Thereadfin bream	Nemipterus spp	1002
	Kerisi bali	Sharptoothed bass	Pristipormoides typus	621
	Kikek	Ponyfish (Slipmounth)	Leiognathus spp.	661
	Merah	Red snapper	Lutianus argentimaculatus/	321
	merah tanda	Redspot	Lutjanus spp.	281
	Delah	Red belly fusiller	Lutjanus spp.	40
	Ketambak	Long face emperor	Lutjanus spp.	29
	Pogot/barat barat	moses perch	Acanthurida spp.	33
	sapariding	Catfish eel	Plotosus spp.	20
	Semilang	Catfish	plotosus spp	34
	Beliak mata	Thereadfin big eye	Prianchantus tayenus	821
	Buntal	Pufferfish	tetraodontidae spp	42
	chelek mata	blotched grunt	pomadasys spp.	611
	Ayam laut	leatherjacket	abalistes stellaris	220
	Julang-julang	Halfbeak	Hemirhamphidae	80
	Selangat	Bony bream	Anodontostoma chacunda	581

	Tedungan	parrotfish	Scrus spp	22
	Kitang	Spotted butterfish	Scatopbagus argus	390
	selunsung	Barramundi	Lates calcarifer	184
	Balais	Golden line spinefoot	Siganus spp	166
	lengging	Eel	muraenidae	100
	Belays putih	Rabbit fish	Siganus spp	40
34 Jacks, Mullets, Sauries, Etc.	Alu-alu/Kacang-kacang	Barracuda	Syhyraena jello/S.optusa	140
	bangus	Milkfish	Chanos chanos spp	30
	Belanak	Mullet	Liza spp./yalamugi spp.	601
	Belanak Kedera	Mullet	Seheli spp	38
	Cincaru	Hardtail scad	Megalaspis cordyla	321
	Demudok/Rambai ikan putih	Horse mackerel	Carangoides spp	140
	Gerong-gerong	Golden trevally	Caranx speciosus	100
	Kurau/Senangin/Senohong	Thereadfin	Polynemus spp.	160
	senangin		polynemidae	262
	Mata besar/Lolong	Big eye scad	Selar crymenophthalmus	290
	Pisang-pisang	Rainbow runner	Elagatis bipinnulatus	80
	Selar kuning	Yellow Striped (Travally)	Selaroides leptolepis	238
	Selayang/Curut	Round scad	Decapterus maruadis/ D.macrosoma	230
	Talang	Queenfish/ Leatherskin	Scomberoides commersonianus	142
35 Herrings, Sardines, Anchovies, Etc.	Parang-parang	Dorab wolf-herring	Chirocentrus dorad	240
	Bulan-bulan	Indo-Pacific torpon	Megalops cyprinoides	123
	Aya/Kayu/Tongkol hitam	Longtail tuna	Thunnus tonggol	367
36 Tunas, Bonitos Billfishes, Etc	Tenggiri	Spanish mackerel	Scomberomorus spp.	549
37 Mackerels, Snoeks, Cutlass, Fishes, Etc.	Kembong /Rumahan	Indian mackerel	Rastrelligerkanagurta spp.	264
	Rrumahan	Club mackerel	kanagarta spp	228
	Kembong	Club mackerel	brachysomsa spp	220
	Tulai	Japanese	Scomber	232

			mackerel	australasicus	
38	Sharks, Rays, Chimaeras, Etc.	Yu Pari	Shark Ray	Galeorhinidae Gymnura spp.	222 320
39	Miscellaneous marine fishes	Ketam suri		Portunus pelagicus	190
		Udang harimau		Monodon spp	461
		u.bunga		penaeid	403
		u.lipan		Panulirus spp.	60
		u.putih		penaeid	120
		sotong katak		Toradodes spp	280
		s.kurita		Toradodes spp	240
		unidentify		Mixed spp.	968
Total					20033

5.3.3 The Catch Ratio of Marketable with Underutilized

A study on catch ratio composition of marketable and underutilized had been carried out in Darvel Bay. During the survey, data were collected using trawler boat. All catch were weighed and divided into groups of marketable and non-marketable based on immature and under size. Non-marketable were the species that cannot be consumed locally or accepted by market while marketable having a market value. There were 128 hauls made in this survey. It was found a total catch of 20,032.9 kilograms while the trash fish was dominant in this survey with 12,426.9 kilograms or 62.03%. The marketable fish were 7,606 kilograms or 37.96% as in Table 5.3. The ratio of marketable fish was lower when compare to unmarketable or trash fish. The estimation of biomass of demersal resources along the West Coast of Sabah in 1995 was about 21% trash fish and 79% commercial fish (Biusing, 1996) in coastal fishing zone.

Table 5.3 Ratio composition marketable and underutilized with 128 hauls

No	Position	Marketable	Underutilized	Total weight
1	Long 118° 24' E, Lat 4° 50' N	50	90	140
2	Long 118° 14' E, Lat 4° 53' N	47	95	142
3	Long 118° 25' E, Lat 4° 58' N	61	120	181

4	Long 118° 15' E, Lat 4° 53' N	60	110	170
5	Long 118° 26' E, Lat 4° 57' N	67	115	182
6	Long 118° 16' E, Lat 4° 52' N	81	135	216
7	Long 118° 27' E, Lat 4° 57' N	83	125.4	208.4
8	Long 118° 15' E, Lat 4° 52' N	86	130.5	216.5
9	Long 118° 25' E, Lat 4° 57' N	73	98	171
10	Long 118° 16' E, Lat 4° 52' N	76	106	182
11	Long 118° 26' E, Lat 4° 56' N	65	120	185
12	Long 118° 14' E, Lat 4° 51' N	70	130	200
13	Long 118° 12' E, Lat 4° 55' N	63	110	173
14	Long 118° 13' E, Lat 4° 47' N	58	103	161
15	Long 118° 14' E, Lat 4° 48' N	45	89	134
16	Long 118° 15' E, Lat 4° 54' N	50	96	146
17	Long 118° 14' E, Lat 4° 47' N	58	90	148
18	Long 118° 13' E, Lat 4° 53' N	49	103	152
19	Long 118° 16' E, Lat 4° 46' N	40	85	125
20	Long 118° 12' E, Lat 4° 54' N	43	95	138
21	Long 118° 18' E, Lat 4° 48' N	47	72	119
22	Long 118° 11' E, Lat 4° 53' N	54	91	145
23	Long 118° 24' E, Lat 4° 55' N	47	90	137
24	Long 118° 33' E, Lat 4° 57' N	57	85	142
25	Long 118° 26' E, Lat 4° 59' N	54	73	127
26	Long 118° 32' E, Lat 4° 52' N	65	85	150
27	Long 118° 23' E, Lat 4° 56' N	67	91	158
28	Long 118° 35' E, Lat 4° 53' N	61	88	149
29	Long 118° 53' E, Lat 4° 23' N	46	69	115
30	Long 118° 34' E, Lat 4° 52' N	55	80	135
31	Long 118° 23' E, Lat 4° 53' N	61	95	156
32	Long 118° 32' E, Lat 4° 52' N	42	80	122
33	Long 118° 23' E, Lat 4° 58' N	55	69	124
34	Long 118° 30' E, Lat 4° 53' N	34	61	95
35	Long 118° 23' E, Lat 4° 57' N	68	75	143
36	Long 118° 30' E, Lat 4° 52' N	35	73	108
37	Long 118° 23' E, Lat 4° 56' N	87	130	217
38	Long 118° 30' E, Lat 4° 54' N	91	130	221
39	Long 118° 23' E, Lat 4° 55' N	81	120	201
40	Long 118° 30' E, Lat 4° 55' N	86	140	226
41	Long 118° 23' E, Lat 4° 54' N	64	90	154
42	Long 118° 31' E, Lat 4° 55' N	67	98	165
43	Long 118° 22' E, Lat 4° 57' N	65	122	187
44	Long 118° 14' E, Lat 4° 51' N	66	133	199
45	Long 118° 22' E, Lat 4° 56' N	91	120	211
46	Long 118° 13' E, Lat 4° 50' N	87	105	192
47	Long 118° 22' E, Lat 4° 56' N	65	90	155
48	Long 118° 15' E, Lat 4° 54' N	68	97	165
49	Long 118° 22' E, Lat 4° 52' N	60	99	159
50	Long 118° 13' E, Lat 4° 53' N	65	105	170
51	Long 118° 22' E, Lat 4° 53' N	55	86	141
52	Long 118° 12' E, Lat 4° 54' N	64	97	161
53	Long 118° 21' E, Lat 4° 57' N	55	73	128
54	Long 118° 14' E, Lat 4° 55' N	65	85	150
55	Long 118° 21' E, Lat 4° 56' N	60	93	153
56	Long 118° 14' E, Lat 4° 57' N	55	86	141
57	Long 118° 21' E, Lat 4° 55' N	55	75	130
58	Long 118° 14' E, Lat 4° 52' N	64	87	151
59	Long 118° 21' E, Lat 4° 54' N	70	99	169
60	Long 118° 13' E, Lat 4° 53' N	65	93	158
61	Long 118° 21' E, Lat 4° 53' N	45	72	117
62	Long 118° 14' E, Lat 4° 53' N	52	76	128

63	Long 118° 20' E, Lat 4° 53' N	60	90	150
64	Long 118° 14' E, Lat 4° 52' N	40	86	126
65	Long 118° 23' E, Lat 4° 57' N	55	68	123
66	Long 118° 14' E, Lat 4° 53' N	50	65	115
67	Long 118° 23' E, Lat 4° 58' N	65	76	141
68	Long 118° 15' E, Lat 4° 53' N	60	73	133
69	Long 118° 23' E, Lat 4° 56' N	80	120	200
70	Long 118° 16' E, Lat 4° 52' N	101	130	231
71	Long 118° 23' E, Lat 4° 55' N	80	130	210
72	Long 118° 16' E, Lat 4° 53' N	107	140	247
73	Long 118° 23' E, Lat 4° 54' N	70	90	160
74	Long 118° 16' E, Lat 4° 52' N	73	98	171
75	Long 118° 22' E, Lat 4° 56' N	90	130	220
76	Long 118° 14' E, Lat 4° 55' N	100	140	240
77	Long 118° 12' E, Lat 4° 55' N	88	115	203
78	Long 118° 13' E, Lat 4° 47' N	80	110	190
79	Long 118° 14' E, Lat 4° 48' N	65	90	155
80	Long 118° 12' E, Lat 4° 54' N	70	98	168
81	Long 118° 14' E, Lat 4° 47' N	60	93	153
82	Long 118° 13' E, Lat 4° 53' N	65	110	175
83	Long 118° 16' E, Lat 4° 46' N	60	90	150
84	Long 118° 12' E, Lat 4° 53' N	63	99	162
85	Long 118° 18' E, Lat 4° 48' N	58	73	131
86	Long 118° 11' E, Lat 4° 53' N	69	90	159
87	Long 118° 24' E, Lat 4° 55' N	60	95	155
88	Long 118° 13' E, Lat 4° 57' N	51	97	148
89	Long 118° 26' E, Lat 4° 59' N	32	75	107
90	Long 118° 13' E, Lat 4° 52' N	46	90	136
91	Long 118° 23' E, Lat 4° 56' N	45	95	140
92	Long 118° 14' E, Lat 4° 53' N	52	90	142
93	Long 118° 53' E, Lat 4° 23' N	55	80	135
94	Long 118° 14' E, Lat 4° 52' N	58	90	148
95	Long 118° 23' E, Lat 4° 53' N	55	93	148
96	Long 118° 13' E, Lat 4° 52' N	40	85	125
97	Long 118° 25' E, Lat 4° 56' N	40	94	134
98	Long 118° 14' E, Lat 4° 55' N	38	80	118
99	Long 118° 24' E, Lat 4° 54' N	40	75	115
100	Long 118° 15' E, Lat 4° 56' N	43	100	143
101	Long 118° 27' E, Lat 4° 54' N	41	95	136
102	Long 118° 16' E, Lat 4° 56' N	70	130	200
103	Long 118° 26' E, Lat 4° 56' N	51	110	161
104	Long 118° 15' E, Lat 4° 57' N	58	120	178
105	Long 118° 26' E, Lat 4° 56 N	40	90	130
106	Long 118° 16' E, Lat 4° 51' N	46	97	143
107	Long 118° 25' E, Lat 4° 56' N	54	110	164
108	Long 118° 12' E, Lat 4° 51' N	61	120	181
109	Long 118° 14' E, Lat 4° 50' N	56	120	176
110	Long 118° 14' E, Lat 4° 43' N	74	110	184
111	Long 118° 13' E, Lat 4° 47' N	51	90	141
112	Long 118° 14' E, Lat 4° 47' N	46	98	144
113	Long 118° 15' E, Lat 4° 54' N	44	90	134
114	Long 118° 16' E, Lat 4° 46' N	51	93	144
115	Long 118° 13' E, Lat 4° 53' N	50	95	145
116	Long 118° 18' E, Lat 4° 48' N	42	90	132
117	Long 118° 12' E, Lat 4° 54' N	38	80	118
118	Long 118° 24' E, Lat 4° 55' N	40	90	130
119	Long 118° 11' E, Lat 4° 53' N	60	88	148
120	Long 118° 26' E, Lat 4° 59' N	59	95	154
121	Long 118° 33' E, Lat 4° 57' N	35	80	115

122	Long 118° 23' E, Lat 4° 56' N	48	94	142
123	Long 118° 32' E, Lat 4° 52' N	58	94	152
124	Long 118° 53' E, Lat 4° 55' N	60	113	173
125	Long 118° 35' E, Lat 4° 21' N	45	80	125
126	Long 118° 32' E, Lat 4° 51' N	46	88	134
127	Long 118° 34' E, Lat 4° 54' N	57	110	167
128	Long 118° 31' E, Lat 4° 53' N	40	90	130
Total (kg)		7606	12426.9	20032.9
Average catch per haul		59.4218	97.0851	156.5070
Average catch per hr		14.8554	24.2712	39.1267
STDEV		15.1354	18.0966	30.7242
Variance		229.0804	327.4873	943.9771
Percent		37.9675	62.0324	100

Trawler is a commercial fishing gear and had been introduced to this district in 1975. This fishing gear plays a role in catching demersal fish with high composition of trash fish. This evidence showed that the fisheries were under pressure, catching with high effort fishing gear and low selectivity.

The average catch per hours was 39.12 kg. The averages catch per haul was 156.5 kg while marketable 59.42 kg (38%) and trash fish 97.08 kg (62%). Low selectivity fishing gear caught immature and low quality fish. It is important to review the mesh size in order to improve selectivity and quality of the catch.

5.3.4 The Demersal Fish Resources

5.3.4.1 Swept Area Method

The stock resources can be assessed by using Swept Area Method within the bay area 834.55 kilometers square and the result estimated in Table 5.4.

Table 5.4 The Result From Sweep Area Method

Total area	834.55	N km sq
Average catch	156.50	Kg per haul
Estimated resources	1034979.15	Kg
Density	1242.16	Kg per square kg
Effective width (W)	0.1393	N km Sq

The study of the demersal resources by using sweep area method in a total area 834.55 N Km. Sq had resulted the potential resources 1034979.15 kilograms with assumption that the species was totally vulnerable where in one vulnerability the most conservative estimate of stock biomass (King, 1996). Besides, the estimated density of the area where the potential resources for every one kilometer square 1242.16 kilograms or 1.2 metric tones per kilometer square. In comparison with the estimated biomass of demersal resources along the West Coast of Sabah in 1972, the fish density was about 4.97 metric tones per kilometer square by KK Penyelidik (Mohd. Shaari *et al*, 1972) 7.74 metric tones per kilometer square in 1987 by RV Rastrelliger (Anonymous, 1989) and 3.14 metric tones per kilometer square in 1993 by KK Manchong (Biusing, 1996).

The survey was basically to estimate the possible populations of overall status of the resources after 23 years of trawling in Darvel Bay. The survey based on experimental trial that could provide an understanding status of resources, in terms of removal resources by trawler. In attempting to estimate the population status, it is also necessary to study the changes in the species composition of the fishery resources.

In lightly trawled, the annual removal of fishery resources may have been low. However, in the most intensive trawled the fishery resources may be removed each year. In high-effort of trawler trawling removes smaller size than actually target size fish. The trawlers caught smaller size fish with less value. The removal of fishery resources is subjected to the density of the resources in that area.

HAMID AWONG FISHERIES MODEL (HAFM) ASSEMENT OF THE THREADFIN BIG EYE (*Priacanthus tayenus*) FISHERY IN DARVEL, SABAH, MALAYSIA

6.1 Introduction

After recruitment, the fish normally grew which eventually increased in weight, at the same time the mortality number of this resource decreased. The total biomass of the stock would increase at maximum level over time, after reaching the maximum point the biomass yield decreased. To maximize harvesting the fishermen should carry out proper handling of the resource.

Declined of the parental stock will increase the risk of recruitment failure. The failure of recruitment would then reduce the number of fish available for capture. With understanding of these phenomena, most of the fisheries management based on maintaining the number of parental stocks at safer level i.e. to ensure a sustainable fisheries. Closing season is among the management tools to reduce the catch activity on smaller immature fish. Reducing the effort and catch quota would then reduce catching larger fish.

A good planning will avoid depletion of stock and over fishing at one area. The stock will decrease due to fishing mortality and natural mortality and only can be recovered through recruitment.

The future of Threadfin big eye (*Priacanthus tayenus*) depends on other factors including information of catches, effort, CPUE and other management regime as well as biology information concerning the above species.

The surplus model was developed to determine the level of effort feasible for exploiting the fishery. However, this model could not be used to determine the sensitivity of various input parameters. The sensitivity determination is important to formulate the appropriate strategy of management. Therefore, the biomass models were developed for this purpose.

This model may be used as an aid in understanding complex natural systems and in determining the sensitivity of model to various input parameters. They may also be useful in analyzing fish stocks, assessing the probable outcome of different management strategies, and in communicating the result of fisheries assessment. In fact the model can be used not only to simulate the past history of the fishery by providing a check of the model, but also for future behavior of the fishery under different rates of exploitation.

This model will be written in Excel spread sheet. The model is also used to evaluate the Threadfin big eye fisheries *(Priacanthus tayenus)* and to provide the recommendation for future management of the fisheries. Two parameters were tested in this study such as the number and yield of the fish base on 100 percent within 50 years started from 2003 to 2053.

6.2 MATERIALS AND METHODS

The background of the Threadfin big eye (*Priacanthus tayenus*) fish and its fishery was reviewed through the historical data and biological parameters provided.

6.2.1 Duration and Location of Study Area

The study was conducted from November 2001 to October 2003 in Darvel Bay, Sabah, Malaysia. The locality, specifically in position within the boundaries of 118° 10′ to 118° 40′ East longitude and 5° 0′ to 4° 40′ North latitude. Data analyses were done in the College University of Science and Technology Malaysia, (KUSTEM), Kuala Terangganu, Malaysia. A Global Position System (GPS) (Plate 4.1) device was used to locate the position of the stations. The locations of the study are shown in Figure 4.1.

6.2.2 Samples Collection

The samples were caught in Darvel Bay. A commercial fleet of trawlers with a capacity of 190 Horse Power (HP) was used in this program. *Priacanthus tayenus* were sampled about 1 Kilometer East of Sakar Island. A covered cod end bag with a diamond mesh size of 38 mm was used. A total of 24,653 samples were collected during field sampling (Appendix 9 and Appendix 10).

6.2.3 Biological Parameter
6.2.4 The background and status of the Threadfin Big Eye (*Priachantus tayenus*) fishery

The Threadfin big eye (*Priachantus tayenus*) had a short life, large and fast growth species. The estimated K-value (growth coefficient) was 0.5 per year, L infinity (theoretical maximum length) 25.5 cm, $W=0.0068L^{3525}$ and t_{max} (Theoretical maximum age) 5.8739 years. The Threadfin big eye (*Priachantus tayenus*) was first trawled in 1975 and formed part of commercial catch by trawler vessels. Landing data 1999 to 2005 had been used to compile annual landing data.

The Biomass model was written in Microsoft Excel spread sheet. The model was used to forecast the numbers of the parental stock and the yield of *Priacanthus tayenus* within 50 years started from 2003 until 2052.

6.2.5 Single Sample Paterson Method

Single sample or Paterson method was used for analysis the length frequency distribution of *Priacanthus tayenus* as in Appendix 9.

6.2.6 Ford-Walford Plot

The size distribution data of *Priacanthus tayenus* were use to calculate the asymptotic average maximum body size (L_∞) and growth rate coefficient (K). The growth rate coefficient described the length of time to attain maximum size or the rate at which L approaches (L_∞). The Ford-Walford plot was used to prepare the von Bertalanffy growth curve equation. the value of slope (b), intercept (a) were obtained by contruct graping L_{t+1} against L_t to get the slope and intercept.

6.2.7 Used Formula Calculations

The biological parameters are fitted into growth von Bertalanffy growth curve equation (von Bertalaffy, 1938), as follows;

a. Asymptotic average maximum body size (L_∞)

$$L_\infty = a/(1-b)$$

b. Growth rate coefficient (K).

$$K = -Ln(b)$$

c. Hypnotically age at which the species is zero length (t_0)

$$t_0 = -a/b$$

Intercept (a) and slope (b) were from von Bertalanffy plot regression between $-Ln(1- L_t)/ L_\infty$ VS Time (t).

d. The assuming economic mature age during harvest. (tm)

$$tm = 4 \text{ years}$$

e. The calculated maximum age.

Life span or the time required for fish to reach 95% of the species asymptotic length (t_{max}) (Taylor, 1958 and King, 1995).

$$t_{max} = (-1/K) \ln(1 - (0.95 L_\infty)/ L_\infty)$$

$$tmax = 5.8739 \text{ years}$$

f. Growth in Weight

The relations between length and weight describes as a power function growing allometrically (Sparre. et al., 1989) as;

$$W_t = qLt^b$$

Where,

q = constant determined empirically

W_∞ = Asymptotic weight or weight at L_∞

Weight at age data (W_t)

Where,

$$W_t = W_\infty(1-\exp(-K(t-t_o)))^b$$

g. The assuming age recruitment. (tr)

Tm = 3 years

h. The assuming age first capture. (tc)

Tm = 4 years

i. Beverton & Holt stock recruitment relationship

Beverton & Holt stock recruitment relationship recruitment (Beverton & Holt, 1957) was predicted by spreadsheet computer program in which parameters and formula given as follow,

Initial stock numbers, SI = 100%

R = S/(a+bS)

Where R = Recruitment and S = Stock

Recruitment, R = aSm exp(-bSm),

Where;

Parameter recruitment a and b constant

Sm is a matured stock.

The number of fish in age class t is given as

At age recruitment tr,

$$Nt+1 = SI*Exp(-M)$$

At age capture tc at 4; in this study age at maturity,

t+1=SI*ExpI-(M+ F)

Age class t, t= I to age t maximum (t_{max})= 5.8739 or 6 years

Age at first capture (tc) = age at first maturity (tm) or at about years 4

Then,

Nt+ I would depend on the age at tm., tc and F and recruitment at the beginning of 4th year was given by Beverton & Holt stock recruitment stock as recruitment Relationship (King,M. 1985).

Mortality Model

The catch or mortality (Pauly 1990, Pauly et, al., 1995) would reverse to normal logarithm to produce a curve fitted with age year against normal logarithm catch of fourth cohort.

The data of this species at the age of 1 to 6 years were collected intensively. The catch data were recorded in 2001 to 2002 to obtain catch sample Z as total mortality with assumption Z constant. Annually Mortality for total mortality (Z), natural mortality (M) and fishing mortality (F).

Length-Converted Catch Curve

$C = (F/Z) N (1 - \exp(-Z)$

Survival Rate

$S = N_t/N_o = \exp(-Zt)$. Where N_t number at time t and N_o is initial number

As 100 or percentages.

Survival = $100 \exp(-Z)$

Mortality

Mortality = $100(1-\exp(-Z))$

Total mortality (Z)

Age – based catch curve.

Cohort curve normal logarithms N against age, the slope (a) equal to total mortality rate (Z)

Natural Mortality (M)

$$M = -\ln(0.01)/t_{max}$$

Where,

M = Instantaneous rate of natural mortality

t_{max} = Life span or the time required for fish to reach 95% of the species asymptotic length

Fishing Mortality (F)

Fishing mortality, $F = 0.04$ per yrs

Fishing mortality, $F = qf$,

$q = 0.000003$, f variables

Assumption with this model;

Recruitment at constant recruitment

$$Nt + 1 = S1 * Exp[-M]$$

Age maturity at Age captures or age maturity at year 4

$$Nt + 1 = S1 * Exp(-M + Z)$$

Thus;

$Nt + 1$ will depends on the age at first maturity or age first capture at year 4

The species information Threadfin big eye fisheries (Priacanthus tayenus) information used in this model available from previous studied in this thesis as;

The biological parameter as are shown in (Table 8.1) The Parameter to be used in this Model and flow chart as shown in Figure 8.1. .

Table 6.1 The Parameter to be used in this Model

von Bertalanffy growth

L_{inf}	25.5 cm
W_{inf} =	350.5 gram
K =	0.51 cm/yrs
t_o =	-0.3434
t_m =	4 yrs
t_{max} =	5.8739
t_c =	4 yrs
t_r =	3yr
W=	$0.0068L^{3.3525}$

Beverton & Holt Relationship

a =	1.3495
b =	0.0003

Mortality

M =	0.784 per yrs
F =	0.039 per yrs
Z =	0.823 per yrs

Fishing Parameters

q =	0.000003
f =	1442
# of boats	1442
Price per kg	RM 5.00

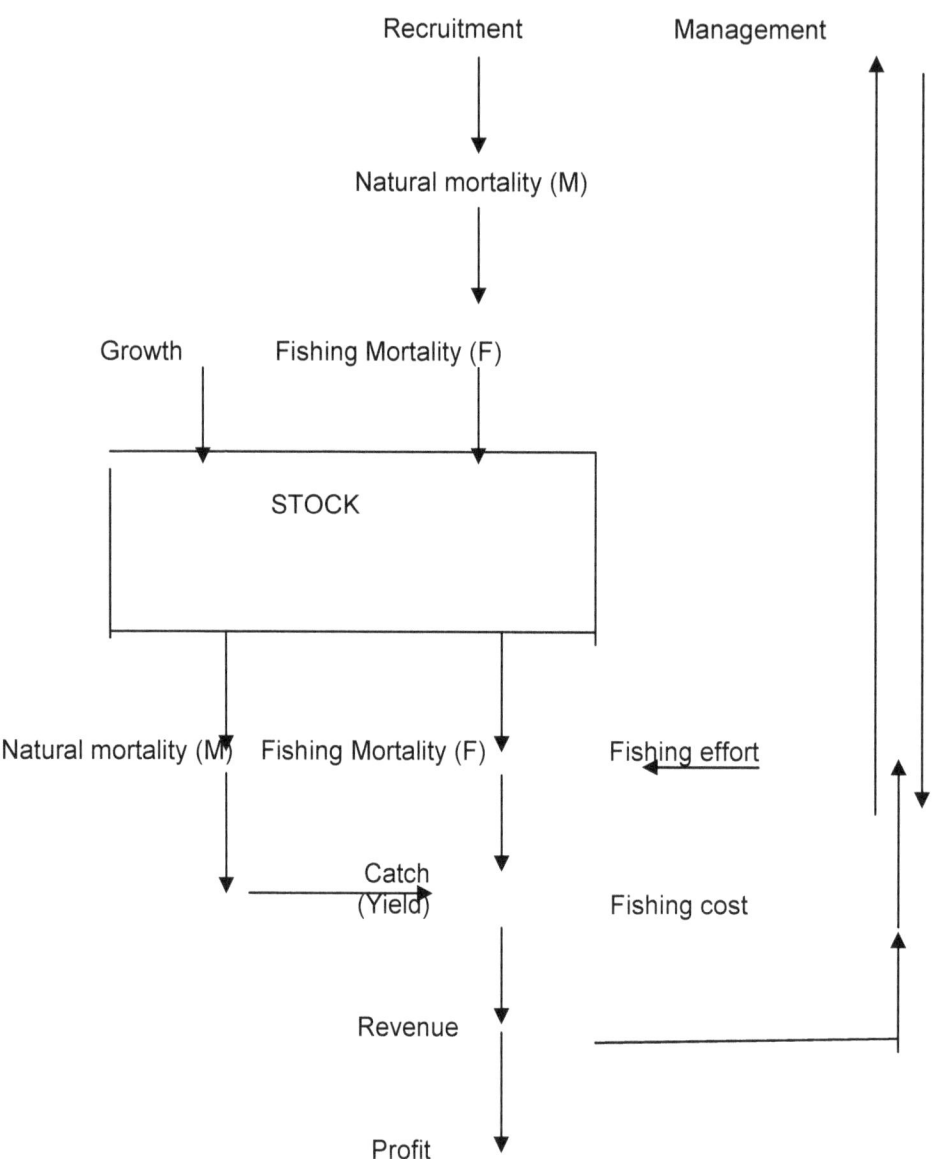

Figure 6.1 Flow chart of the HAFM model for threadfin big eye *(Priacanthus tayenus)* fishery

6.2.8 Stock Structure

The Threadfin big eye (*Priachantus tayenus*) was assumed the type of gear (trawl) could catch uniform on the continental shelf break and along the slope. The catch ability coefficient was 3.0×10^{-6} and estimated from present fishing mortality (F) divided by standard effort (f) where F=qf.

6.2.9 Commercial Fishery

The data was taken from trawlers started in 1991 to 2000 in Sabah, Malaysia. Fishing Parameters for the catching ability coefficient was 3.0×10^{-6} and standard effort of 1442 vessels.

6.3 RESULTS AND DISCUSSION

6.3.1 Beverton & Holt Recruitment

Beverton & Holt recruitment was used to calculate the Parameter recruitment

Regression of stock (year 0+1) between Ln(R/S) year 0+1 (a) = Exp(intercept) = 1.4655, (b) = -(Slope) = 0.0003

Regression of stock (year 0+1+2) between Ln(R/S) year 0+1+2 (a) = Exp(intercept) = 1.349589, b = -(Slope) = 0.0003.

This species have low recruitment only 22.39 % for stock of year 0 and 1, for the year 0,1 and 2 was 3.1%. The recruitment pattern as in Table 6.2 and Table 6.3. Regression of stock (year 0+1+2) between Ln(R/S) Year 0+1+2 Figure 6.2 and Figure 6.3 Regression of stock (year 0+1) between Ln(R/S) year 0+1.

Table 6.2 The species stock

Stock/Year	0	1	2
1	2050	1795	1523
2	1808	1521	1211
3	1675	1459	895
4	1456	0	47

Table 6.3 The stock recruitment in natural logarithms

R o	S(O+1+2	LN (R/S)	S(o+1)	lnR/S (0+1)
2050	5368	-0.96261561	3845	-0.6289338
1808	4540	-0.92070575	3329	-0.6104507
1675	4029	-0.87770504	3134	-0.626497
1456	1503	-0.03177016	1456	0

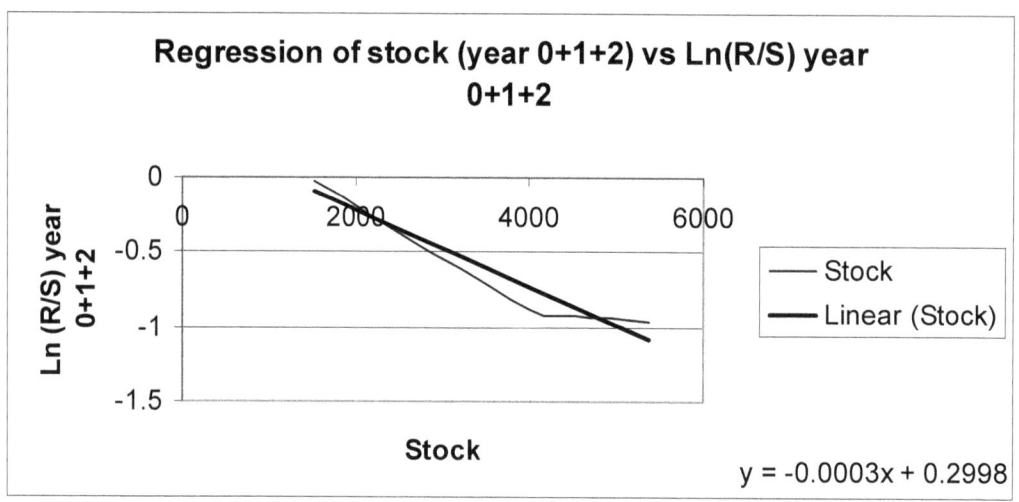

Figure 6.2 Regression of stock (year 0+1+2) between Ln(R/S) Year 0+1+2

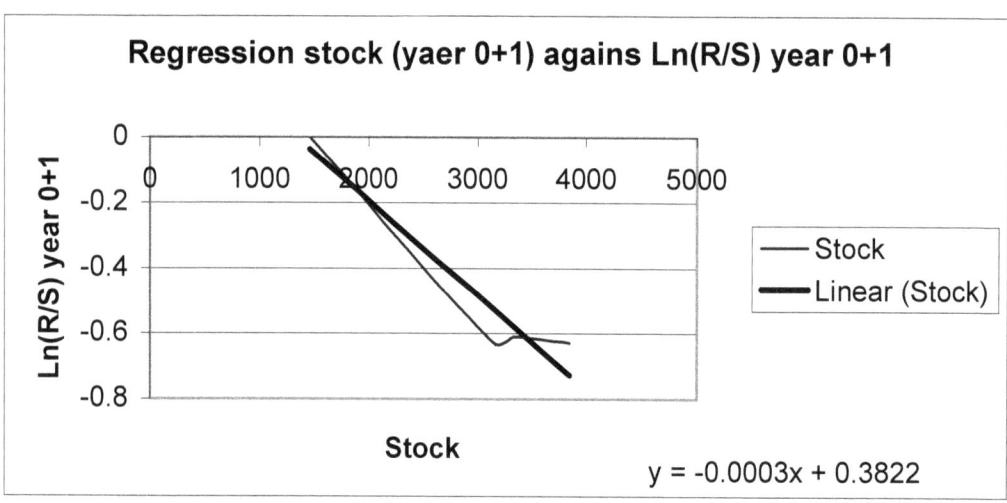

Figure 6.3 Regression of stock (year 0+1) between Ln(R/S) year 0+1

Stock 0+1,   a = exp(intercept) = 1.465505157

   b = -(Slope)   = 0.0003

Stock 0+1+2, a = exp(intercept) = 1.349589

   b = -(Slope)   = 0.0003

Where;

R = Recruitment of the stock

S = Stock of cohort year 0, 1 and 3

The growth of individual to the gain the biomass without sort of information and parameter, the fishery likely to face a risk of growth over fishing a large amount of fish caught when there are still too small for optimum yield as are shown in (Table 6.4 and Figure 6.4) the stock recruitment of the species.

Table 6.4 The stock recruitment of the species

Stock	Rec.(+0+1) R=S/(a+bS)	Rec.(+0+1+2) R=S/(a+bS)
0	0	0
500	309.500714	333.424722
1000	566.410127	606.211658
1500	783.083248	833.523718
2000	968.286133	1025.85732
2500	1128.41082	1190.70931
3000	1268.22805	1333.57701
3500	1391.37063	1458.5832
4000	1500.65363	1568.8804
4500	1598.29222	1666.92049
5000	1686.05338	1754.63909
5500	1765.36379	1833.58462
6000	1837.388	1905.01055
6500	1903.08599	1969.9424
7000	1963.25617	2029.22733
7500	2018.56805	2083.57129
8000	2069.5872	2133.56725
8500	2116.79469	2179.71696
9000	2160.60229	2222.44783
9500	2201.36453	2262.1262
10000	2239.3883	2299.06787
10500	2274.94058	2333.54654
11000	2308.25477	2365.80057
11500	2339.53574	2396.03856
12000	2368.96413	2424.44379
12500	2396.69977	2451.17799
13000	2422.88463	2476.38441
13500	2447.64525	2500.19036
14000	2471.09474	2522.7094

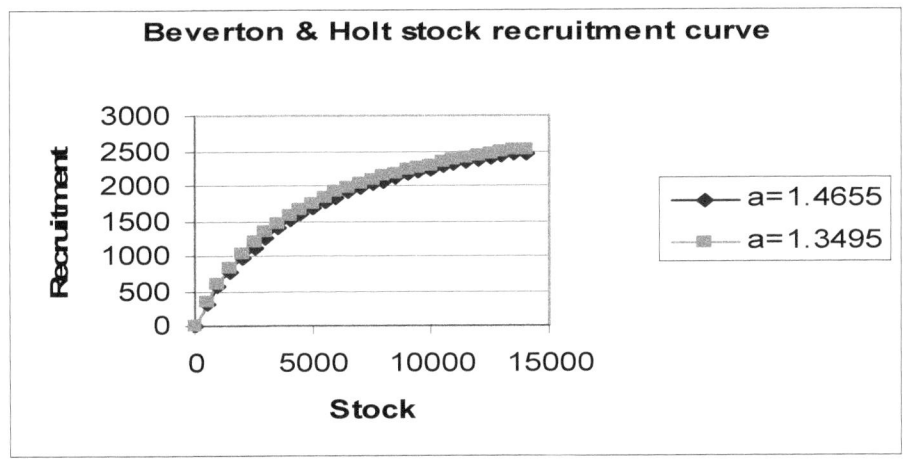

Figure 6.4 The recruitment pattern Beverton & Holt Stock Recruitment

6.3.2 Population Stock (Sm) Model

The assumption of initials number of parental stock in 2003 was 100. After six modes through natural mortality and fishing mortality the parental stock (sm) was 15.8475. In 50 years later or in 2052 the parental stock at first mode would be 2.3 x10^{-8} and through recruitment, natural mortality (M) and Fishing mortality the parental stock 1.7 x10^{-8} as shown in Table 6.5 and Figure 6.5. The fallen of number of parental stock was due to the low recruitment, natural mortality and fishing mortality.

Table 6.5 The population of the parental (Sm) stock *P.tayenus* from 2003 to 2053 (50 years) with initial number stock 100 heads

Mean Wt	33.39825283	104.6716043	178.9129963	237.9173156	279.2523536	306.358785	
Year	1	2	3	4	5	6	Sm
2003	100.00000000000	45.657604962332	20.846168908963	9.5178614502348	4.3456275818102	1.9841094744370	15.847598506482
2004	20.47448950449	45.657604962332	20.846168908963	9.1538099993202	4.1794104084920	1.9082186940638	15.241439101876
2005	20.16310374259	9.348161536014	20.846168908963	9.1538099993202	4.0195509241809	1.8352306822223	15.008591605723
2006	20.06922207403	9.205990254939	4.268146665354	9.1538099993202	4.0195509241809	1.7650344100744	14.938395333575
2007	10.31180523433	9.163126133573	4.203234663471	1.8741958675717	4.0195509241809	1.7650344100744	7.658781201827
2008	5.97533733575	4.708123298377	4.183663932267	1.8456922065629	0.8229825320990	1.7650344100744	4.433709148736
2009	4.05690949394	2.728195915923	2.149616336713	1.8370984569983	0.8104662228290	0.3613817850413	3.008946464869
2010	2.84092771838	1.852287710421	1.245628913891	0.9439230586509	0.8066926013526	0.3558857191958	2.106501379199
2011	1.77482023703	1.297099554925	0.845710205590	0.5469710265329	0.4144882625964	0.3542286754409	1.315687964570
2012	1.07064138542	0.810340412614	0.592224590756	0.3713617869193	0.2401817271020	0.1820069106859	0.793550424707
2013	0.71321832518	0.488829214320	0.369982024442	0.2600531255589	0.1630695430567	0.1054667600940	0.528589428710
2014	0.46993054599	0.325638405428	0.223187711615	0.1624636723271	0.1141925363596	0.0716058485535	0.348262057240
2015	0.29617905811	0.214559032284	0.148678696756	0.0980044781957	0.0713398032401	0.0501433517948	0.219487633231
2016	0.18844682993	0.135228264333	0.097962515371	0.0652866503669	0.0430349757024	0.0313261879005	0.139647813970
2017	0.12223696839	0.086040309174	0.061741986727	0.0430165493084	0.0286681737810	0.0188971888612	0.090581911951
2018	0.07906489047	0.055810472148	0.039283944471	0.0271116682372	0.0188890976042	0.0125885488583	0.058589314700
2019	0.05053762513	0.036099135354	0.025481724901	0.0172500647616	0.0119050680675	0.0082944358401	0.037449568669
2020	0.03237659098	0.023074269240	0.016482000615	0.0111893398358	0.0075747162941	0.0052276622911	0.023991718421
2021	0.02088608542	0.014782376009	0.010535158698	0.0072374498496	0.0049133771928	0.0033261513930	0.015476978435
2022	0.01344326894	0.009536086374	0.006749278842	0.0046261181828	0.0031780535355	0.0021575245540	0.009961696272

HAMID AWONG FISHERIES MODEL (HAFM)

Year							
2023	0.00862409779	0.00613787428	0.00435394856456	0.00296369126223	0.00203138525781	0.00139552205230	0.0063905997234
2024	0.00554005857	0.00393755628502	0.00280240655510	0.00191187245758	0.00130139335618	0.00089200604736	0.0041052726113
2025	0.0035647771453	0.00252945805552	0.00179779399282	0.00123057129686	0.00083952683966	0.00057145797168	0.0026415561080
2026	0.00229204248	0.00162759186866	0.00115488996632	0.00078943352617	0.00054035904115	0.00036864659468	0.0016984391618
2027	0.00147237792	0.00104649170272	0.00074311946441	0.00050712643503	0.00034665000250	0.00023727832275	0.0010910547602
2028	0.00094629173	0.00067225249331	0.00047780304710	0.00032631292650	0.00022268547530	0.00015221829364	0.0007012166954
2029	0.00060846408	0.00043205414125	0.00030693438757	0.00020980921380	0.00014328803260	0.00009778394004	0.0004508811864
2030	0.00039112290	0.00027781012581	0.00019726557277	0.00013477867700	0.00009212981470	0.00006291954320	0.0002898280349
2031	0.00025135794	0.00017857734883	0.00012684144964	0.00008662174711	0.00005918298029	0.00004045533853	0.0001862600658
2032	0.00016156524	0.00011476401710	0.00008153414022	0.00005569764574	0.00003803667806	0.00002598797698	0.0001197223006
2033	0.00010386095	0.00007376681919	0.00005239850175	0.00003580264716	0.00002445752355	0.00001670237472	0.0000769625452
2034	0.00006675951	0.00004742042326	0.00003368016261	0.00002300882874	0.00001572138415	0.00001073960046	0.0000494698131
2035	0.00004290909	0.00003048030480	0.00002165102961	0.00001478937503	0.00001010346064	0.00000690345383	0.0000317962893
2036	0.00002758106	0.00001959126419	0.00001391680061	0.00000950723421	0.00000649419712	0.00000443655421	0.0000204379855
2037	0.00001772901	0.00001259285299	0.00000894490185	0.00000611103896	0.00000417474391	0.00000285168211	0.0000131374648
2038	0.00001139576	0.00000809464117	0.00000574959466	0.00000392781683	0.00000268343264	0.00000183318155	0.0000084444309
2039	0.00000732481	0.00000520303103	0.00000369581918	0.00000252471791	0.00000172475283	0.00000117832842	0.0000054277991
2040	0.00000470824	0.00000334433553	0.00000237557938	0.00000162287986	0.00000110863482	0.00000075736021	0.0000034888748
2041	0.00000302637	0.00000214966827	0.00000152694331	0.00000104314625	0.00000071266526	0.00000048681529	0.0000022425879
2042	0.00000194528	0.00000138176924	0.00000098148694	0.00000067049964	0.00000045805840	0.00000031299231	0.0000014414811
2043	0.00000125037	0.00000088816878	0.00000063088278	0.00000043098309	0.00000029442470	0.00000020112391	0.0000009265468
2044	0.00000080371	0.00000057089151	0.00000040551610	0.00000027702847	0.00000018925000	0.00000012928555	0.0000005955638
2045	0.00000051661	0.00000036696553	0.00000026065539	0.00000017806716	0.00000012164666	0.00000008310206	0.0000003828157
2046	0.00000033206	0.00000023587226	0.00000016754354	0.00000011445694	0.00000007819156	0.00000005341654	0.0000002460649
2047	0.00000021344	0.00000015161367	0.00000010769335	0.00000007357042	0.00000005025957	0.00000003433480	0.0000001581647
2048	0.00000013720	0.00000009745345	0.00000006922228	0.00000004728950	0.00000003230570	0.00000002206955	0.0000001016647
2049	0.00000008819	0.00000006264193	0.00000004449475	0.00000003039666	0.00000002076542	0.00000001418583	0.0000000653478
2050	0.00000005668	0.00000004026484	0.00000002860028	0.00000001953824	0.00000001334751	0.00000000911831	0.0000000420040
2051	0.00000003644	0.00000002588125	0.00000001838368	0.00000001255872	0.00000000857955	0.00000000586111	0.0000000269992
2052	0.00000002342	0.00000001663611	0.00000001181658	0.00000000807252	0.00000000551475	0.00000000376735	0.0000000173545

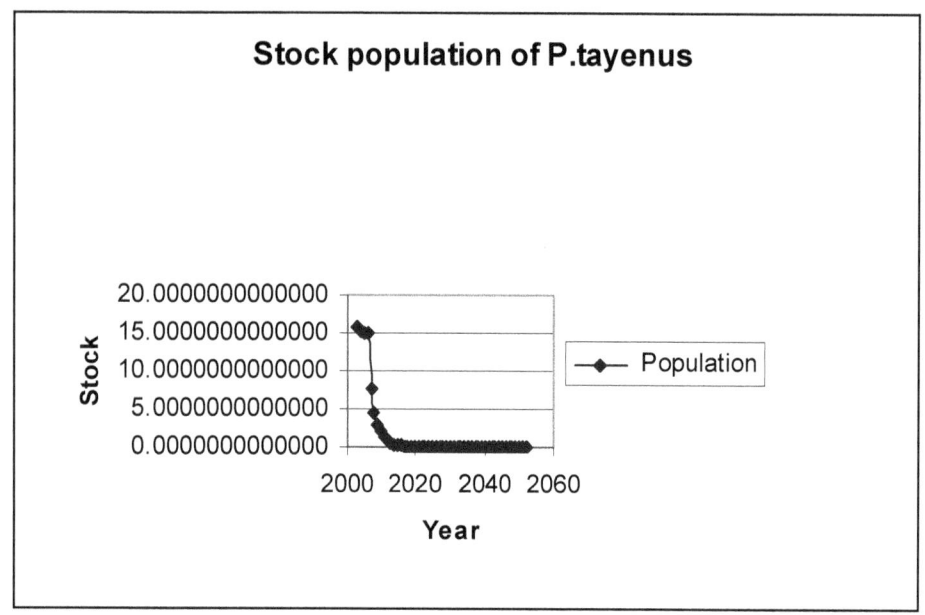

Figure 6.5 The Population of the parental stock *Priacanthus tayenus* from 2003 to 2053 (50 years)

Design Model the Numbers of Parental Stock (Sm)

Mean Weight

Age year 1	Winf*(1-Exp(-K*(yr1+to)))^b
Age year 2	Winf*(1-Exp(-K*(yr1+to)))^b
Age year 3	Winf*(1-Exp(-K*(yr1+to)))^b
Age year 4	Winf*(1-Exp(-K*(yr1+to)))^b
Age year 5	Winf*(1-Exp(-K*(yr1+to)))^b
Age year 6	Winf*(1-Exp(-K*(yr1+to)))^b

Year 2003

1

Pop. Age year 1	initial (100)
Pop. Age year 2	Pop. Age year 1*Exp(-M)
Pop. Age year 3	Pop. Age year 2*Exp(-M)
Pop. Age year 4	Pop. Age year 3*Exp(-M)
Pop. Age year 5	Pop. Age year 4*Exp(-M)
Pop. Age year 6	Pop. Age year 5*Exp(-M)
Total Sm stock	Sum(Pop.Age yr1:yr6)

2   Year 2004

Pop. Age year 1	(R B&H a Total Sm Yr2)*Exp(R B&H b*Total Sm Yr2)
Pop. Age year 2	Pop. Age year 1*Exp(-M)
Pop. Age year 3	Pop. Age year 2*Exp(-M)
Pop. Age year 4	Pop. Age year 3*Exp(-(Z))
Pop. Age year 5	Pop. Age year 4*Exp(-(Z))
Pop. Age year 6	Pop. Age year 5*Exp(-(Z))
Total Sm stock	Sum(Pop.Age yr1:yr6)

3   Year 2005

HAMID AWONG FISHERIES MODEL (HAFM)

		Pop. Age year 1	(R B&H a Total Sm Yr3)*Exp(R B&H b*Total Sm Yr3)
		Pop. Age year 2	Pop. Age year 1*Exp(-M)
		Pop. Age year 3	Pop. Age year 2*Exp(-M)
		Pop. Age year 4	Pop. Age year 3*Exp(-(Z))
		Pop. Age year 5	Pop. Age year 4*Exp(-(Z))
		Pop. Age year 6	Pop. Age year 5*Exp(-(Z))
		Total Sm stock	Sum(Pop.Age yr1:yr6)
4	**Year 2006**		
		Pop. Age year 1	(R B&H a Total Sm Yr4)*Exp(R B&H b*Total Sm Yr4)
		Pop. Age year 2	Pop. Age year 1*Exp(-M)
		Pop. Age year 3	Pop. Age year 2*Exp(-M)
		Pop. Age year 4	Pop. Age year 3*Exp(-(Z))
		Pop. Age year 5	Pop. Age year 4*Exp(-(Z))
		Pop. Age year 6	Pop. Age year 5*Exp(-(Z))
		Total Sm stock	Sum(Pop.Age yr1:yr6)
5	**Year 2007**		
		Pop. Age year 1	(R B&H a Total Sm Yr5)*Exp(R B&H b*Total Sm Yr5)
		Pop. Age year 2	Pop. Age year 1*Exp(-M)
		Pop. Age year 3	Pop. Age year 2*Exp(-M)
		Pop. Age year 4	Pop. Age year 3*Exp(-(Z))
		Pop. Age year 5	Pop. Age year 4*Exp(-(Z))
		Pop. Age year 6	Pop. Age year 5*Exp(-(Z))
		Total Sm stock	Sum(Pop.Age yr1:yr6)
6	**Year 2008**		
		Pop. Age year 1	(R B&H a Total Sm Yr6)*Exp(R B&H b*Total Sm Yr6)
		Pop. Age year 2	Pop. Age year 1*Exp(-M)
		Pop. Age year 3	Pop. Age year 2*Exp(-M)
		Pop. Age year 4	Pop. Age year 3*Exp(-(Z))
		Pop. Age year 5	Pop. Age year 4*Exp(-(Z))
		Pop. Age year 6	Pop. Age year 5*Exp(-(Z))
		Total Sm stock	Sum(Pop.Age yr1:yr6)
7	**Year 2009**		
		Pop. Age year 1	(R B&H a Total Sm Yr7)*Exp(R B&H b*Total Sm Yr7)
		Pop. Age year 2	Pop. Age year 1*Exp(-M)
		Pop. Age year 3	Pop. Age year 2*Exp(-M)
		Pop. Age year 4	Pop. Age year 3*Exp(-(Z))
		Pop. Age year 5	Pop. Age year 4*Exp(-(Z))
		Pop. Age year 6	Pop. Age year 5*Exp(-(Z))
		Total Sm stock	Sum(Pop.Age yr1:yr6)
8	**Year 2010**		
		Pop. Age year 1	(R B&H a Total Sm Yr8)*Exp(R B&H b*Total Sm Yr8)
		Pop. Age year 2	Pop. Age year 1*Exp(-M)
		Pop. Age year 3	Pop. Age year 2*Exp(-M)
		Pop. Age year 4	Pop. Age year 3*Exp(-(Z))
		Pop. Age year 5	Pop. Age year 4*Exp(-(Z))
		Pop. Age year 6	Pop. Age year 5*Exp(-(Z))
		Total Sm stock	Sum(Pop.Age yr1:yr6)
9	**Year 2011**		
		Pop. Age year 1	(R B&H a Total Sm Yr9)*Exp(R B&H b*Total Sm Yr9)
		Pop. Age year 2	Pop. Age year 1*Exp(-M)
		Pop. Age year 3	Pop. Age year 2*Exp(-M)
		Pop. Age year 4	Pop. Age year 3*Exp(-(Z))
		Pop. Age year 5	Pop. Age year 4*Exp(-(Z))

		Pop. Age year 6	Pop. Age year 5*Exp(-(Z))
		Total Sm stock	Sum(Pop.Age yr1:yr6)

10   Year 2012

Pop. Age year 1	(R B&H a Total Sm Yr10)*Exp(R B&H b*Total Sm Yr10)
Pop. Age year 2	Pop. Age year 1*Exp(-M)
Pop. Age year 3	Pop. Age year 2*Exp(-M)
Pop. Age year 4	Pop. Age year 3*Exp(-(Z))
Pop. Age year 5	Pop. Age year 4*Exp(-(Z))
Pop. Age year 6	Pop. Age year 5*Exp(-(Z))
Total Sm stock	Sum(Pop.Age yr1:yr6)

11   Year 2013

Pop. Age year 1	(R B&H a Total Sm Yr11)*Exp(R B&H b*Total Sm Yr11)
Pop. Age year 2	Pop. Age year 1*Exp(-M)
Pop. Age year 3	Pop. Age year 2*Exp(-M)
Pop. Age year 4	Pop. Age year 3*Exp(-(Z))
Pop. Age year 5	Pop. Age year 4*Exp(-(Z))
Pop. Age year 6	Pop. Age year 5*Exp(-(Z))
Total Sm stock	Sum(Pop.Age yr1:yr6)

12   Year 2014

Pop. Age year 1	(R B&H a Total Sm Yr12)*Exp(R B&H b*Total Sm Yr12)
Pop. Age year 2	Pop. Age year 1*Exp(-M)
Pop. Age year 3	Pop. Age year 2*Exp(-M)
Pop. Age year 4	Pop. Age year 3*Exp(-(Z))
Pop. Age year 5	Pop. Age year 4*Exp(-(Z))
Pop. Age year 6	Pop. Age year 5*Exp(-(Z))
Total Sm stock	Sum(Pop.Age yr1:yr6)

13   Year 2015

Pop. Age year 1	(R B&H a Total Sm Yr13)*Exp(R B&H b*Total Sm Yr13)
Pop. Age year 2	Pop. Age year 1*Exp(-M)
Pop. Age year 3	Pop. Age year 2*Exp(-M)
Pop. Age year 4	Pop. Age year 3*Exp(-(Z))
Pop. Age year 5	Pop. Age year 4*Exp(-(Z))
Pop. Age year 6	Pop. Age year 5*Exp(-(Z))
Total Sm stock	Sum(Pop.Age yr1:yr6)

14   Year 2016

Pop. Age year 1	(R B&H a Total Sm Yr14)*Exp(R B&H b*Total Sm Yr14)
Pop. Age year 2	Pop. Age year 1*Exp(-M)
Pop. Age year 3	Pop. Age year 2*Exp(-M)
Pop. Age year 4	Pop. Age year 3*Exp(-(Z))
Pop. Age year 5	Pop. Age year 4*Exp(-(Z))
Pop. Age year 6	Pop. Age year 5*Exp(-(Z))
Total Sm stock	Sum(Pop.Age yr1:yr6)

15   Year 2017

Pop. Age year 1	(R B&H a Total Sm Yr15)*Exp(R B&H b*Total Sm Yr15)
Pop. Age year 2	Pop. Age year 1*Exp(-M)
Pop. Age year 3	Pop. Age year 2*Exp(-M)
Pop. Age year 4	Pop. Age year 3*Exp(-(Z))
Pop. Age year 5	Pop. Age year 4*Exp(-(Z))
Pop. Age year 6	Pop. Age year 5*Exp(-(Z))
Total Sm stock	Sum(Pop.Age yr1:yr6)

16   Year 2018

Pop. Age year 1	(R B&H a Total Sm Yr16)*Exp(R B&H b*Total Sm Yr16)

		Pop. Age year 2	Pop. Age year 1*Exp(-M)
		Pop. Age year 3	Pop. Age year 2*Exp(-M)
		Pop. Age year 4	Pop. Age year 3*Exp(-(Z))
		Pop. Age year 5	Pop. Age year 4*Exp(-(Z))
		Pop. Age year 6	Pop. Age year 5*Exp(-(Z))
		Total Sm stock	Sum(Pop.Age yr1:yr6)
17	**Year 2019**		
		Pop. Age year 1	(R B&H a Total Sm Yr17)*Exp(R B&H b*Total Sm Yr17)
		Pop. Age year 2	Pop. Age year 1*Exp(-M)
		Pop. Age year 3	Pop. Age year 2*Exp(-M)
		Pop. Age year 4	Pop. Age year 3*Exp(-(Z))
		Pop. Age year 5	Pop. Age year 4*Exp(-(Z))
		Pop. Age year 6	Pop. Age year 5*Exp(-(Z))
		Total Sm stock	Sum(Pop.Age yr1:yr6)
18	**Year 2020**		
		Pop. Age year 1	(R B&H a Total Sm Yr18)*Exp(R B&H b*Total Sm Yr18)
		Pop. Age year 2	Pop. Age year 1*Exp(-M)
		Pop. Age year 3	Pop. Age year 2*Exp(-M)
		Pop. Age year 4	Pop. Age year 3*Exp(-(Z))
		Pop. Age year 5	Pop. Age year 4*Exp(-(Z))
		Pop. Age year 6	Pop. Age year 5*Exp(-(Z))
		Total Sm stock	Sum(Pop.Age yr1:yr6)
19	**Year 2021**		
		Pop. Age year 1	(R B&H a Total Sm Yr19)*Exp(R B&H b*Total Sm Yr19)
		Pop. Age year 2	Pop. Age year 1*Exp(-M)
		Pop. Age year 3	Pop. Age year 2*Exp(-M)
		Pop. Age year 4	Pop. Age year 3*Exp(-(Z))
		Pop. Age year 5	Pop. Age year 4*Exp(-(Z))
		Pop. Age year 6	Pop. Age year 5*Exp(-(Z))
		Total Sm stock	Sum(Pop.Age yr1:yr6)
20	**Year 2022**		
		Pop. Age year 1	(R B&H a Total Sm Yr20)*Exp(R B&H b*Total Sm Yr20)
		Pop. Age year 2	Pop. Age year 1*Exp(-M)
		Pop. Age year 3	Pop. Age year 2*Exp(-M)
		Pop. Age year 4	Pop. Age year 3*Exp(-(Z))
		Pop. Age year 5	Pop. Age year 4*Exp(-(Z))
		Pop. Age year 6	Pop. Age year 5*Exp(-(Z))
		Total Sm stock	Sum(Pop.Age yr1:yr6)
21	**Year 2023**		
		Pop. Age year 1	(R B&H a Total Sm Yr21)*Exp(R B&H b*Total Sm Yr21)
		Pop. Age year 2	Pop. Age year 1*Exp(-M)
		Pop. Age year 3	Pop. Age year 2*Exp(-M)
		Pop. Age year 4	Pop. Age year 3*Exp(-(Z))
		Pop. Age year 5	Pop. Age year 4*Exp(-(Z))
		Pop. Age year 6	Pop. Age year 5*Exp(-(Z))
		Total Sm stock	Sum(Pop.Age yr1:yr6)
22	**Year 2024**		
		Pop. Age year 1	(R B&H a Total Sm Yr22)*Exp(R B&H b*Total Sm Yr22)
		Pop. Age year 2	Pop. Age year 1*Exp(-M)
		Pop. Age year 3	Pop. Age year 2*Exp(-M)
		Pop. Age year 4	Pop. Age year 3*Exp(-(Z))
		Pop. Age year 5	Pop. Age year 4*Exp(-(Z))
		Pop. Age year 6	Pop. Age year 5*Exp(-(Z))

		Total Sm stock	Sum(Pop.Age yr1:yr6)
23	**Year 2025**		
		Pop. Age year 1	(R B&H a Total Sm Yr23)*Exp(R B&H b*Total Sm Yr23)
		Pop. Age year 2	Pop. Age year 1*Exp(-M)
		Pop. Age year 3	Pop. Age year 2*Exp(-M)
		Pop. Age year 4	Pop. Age year 3*Exp(-(Z))
		Pop. Age year 5	Pop. Age year 4*Exp(-(Z))
		Pop. Age year 6	Pop. Age year 5*Exp(-(Z))
		Total Sm stock	Sum(Pop.Age yr1:yr6)
24	**Year 2026**		
		Pop. Age year 1	(R B&H a Total Sm Yr24)*Exp(R B&H b*Total Sm Yr24)
		Pop. Age year 2	Pop. Age year 1*Exp(-M)
		Pop. Age year 3	Pop. Age year 2*Exp(-M)
		Pop. Age year 4	Pop. Age year 3*Exp(-(Z))
		Pop. Age year 5	Pop. Age year 4*Exp(-(Z))
		Pop. Age year 6	Pop. Age year 5*Exp(-(Z))
		Total Sm stock	Sum(Pop.Age yr1:yr6)
25	**Year 2027**		
		Pop. Age year 1	(R B&H a Total Sm Yr25)*Exp(R B&H b*Total Sm Yr25)
		Pop. Age year 2	Pop. Age year 1*Exp(-M)
		Pop. Age year 3	Pop. Age year 2*Exp(-M)
		Pop. Age year 4	Pop. Age year 3*Exp(-(Z))
		Pop. Age year 5	Pop. Age year 4*Exp(-(Z))
		Pop. Age year 6	Pop. Age year 5*Exp(-(Z))
		Total Sm stock	Sum(Pop.Age yr1:yr6)
26	**Year 2028**		
		Pop. Age year 1	(R B&H a Total Sm Yr26)*Exp(R B&H b*Total Sm Yr26)
		Pop. Age year 2	Pop. Age year 1*Exp(-M)
		Pop. Age year 3	Pop. Age year 2*Exp(-M)
		Pop. Age year 4	Pop. Age year 3*Exp(-(Z))
		Pop. Age year 5	Pop. Age year 4*Exp(-(Z))
		Pop. Age year 6	Pop. Age year 5*Exp(-(Z))
		Total Sm stock	Sum(Pop.Age yr1:yr6)
27	**Year 2029**		
		Pop. Age year 1	(R B&H a Total Sm Yr27)*Exp(R B&H b*Total Sm Yr27)
		Pop. Age year 2	Pop. Age year 1*Exp(-M)
		Pop. Age year 3	Pop. Age year 2*Exp(-M)
		Pop. Age year 4	Pop. Age year 3*Exp(-(Z))
		Pop. Age year 5	Pop. Age year 4*Exp(-(Z))
		Pop. Age year 6	Pop. Age year 5*Exp(-(Z))
		Total Sm stock	Sum(Pop.Age yr1:yr6)
28	**Year 2030**		
		Pop. Age year 1	(R B&H a Total Sm Yr28)*Exp(R B&H b*Total Sm Yr28)
		Pop. Age year 2	Pop. Age year 1*Exp(-M)
		Pop. Age year 3	Pop. Age year 2*Exp(-M)
		Pop. Age year 4	Pop. Age year 3*Exp(-(Z))
		Pop. Age year 5	Pop. Age year 4*Exp(-(Z))
		Pop. Age year 6	Pop. Age year 5*Exp(-(Z))
		Total Sm stock	Sum(Pop.Age yr1:yr6)
29	**Year 2031**		
		Pop. Age year 1	(R B&H a Total Sm Yr29)*Exp(R B&H b*Total Sm Yr29)
		Pop. Age year 2	Pop. Age year 1*Exp(-M)

HAMID AWONG FISHERIES MODEL (HAFM)

		Pop. Age year 3	Pop. Age year 2*Exp(-M)
		Pop. Age year 4	Pop. Age year 3*Exp(-(Z))
		Pop. Age year 5	Pop. Age year 4*Exp(-(Z))
		Pop. Age year 6	Pop. Age year 5*Exp(-(Z))
		Total Sm stock	Sum(Pop.Age yr1:yr6)
30	**Year 2032**		
		Pop. Age year 1	(R B&H a Total Sm Yr30)*Exp(R B&H b*Total Sm Yr30)
		Pop. Age year 2	Pop. Age year 1*Exp(-M)
		Pop. Age year 3	Pop. Age year 2*Exp(-M)
		Pop. Age year 4	Pop. Age year 3*Exp(-(Z))
		Pop. Age year 5	Pop. Age year 4*Exp(-(Z))
		Pop. Age year 6	Pop. Age year 5*Exp(-(Z))
		Total Sm stock	Sum(Pop.Age yr1:yr6)
31	**Year 2033**		
		Pop. Age year 1	(R B&H a Total Sm Yr31)*Exp(R B&H b*Total Sm Yr31)
		Pop. Age year 2	Pop. Age year 1*Exp(-M)
		Pop. Age year 3	Pop. Age year 2*Exp(-M)
		Pop. Age year 4	Pop. Age year 3*Exp(-M)
		Pop. Age year 5	Pop. Age year 4*Exp(-M)
		Pop. Age year 6	Pop. Age year 5*Exp(-M)
		Total Sm stock	Sum(Pop.Age yr1:yr6)
32	**Year 2034**		
		Pop. Age year 1	(R B&H a Total Sm Yr32)*Exp(R B&H b*Total Sm Yr32)
		Pop. Age year 2	Pop. Age year 1*Exp(-M)
		Pop. Age year 3	Pop. Age year 2*Exp(-M)
		Pop. Age year 4	Pop. Age year 3*Exp(-(Z))
		Pop. Age year 5	Pop. Age year 4*Exp(-(Z))
		Pop. Age year 6	Pop. Age year 5*Exp(-(Z))
		Total Sm stock	Sum(Pop.Age yr1:yr6)
33	**Year 2035**		
		Pop. Age year 1	(R B&H a Total Sm Yr33)*Exp(R B&H b*Total Sm Yr33)
		Pop. Age year 2	Pop. Age year 1*Exp(-M)
		Pop. Age year 3	Pop. Age year 2*Exp(-M)
		Pop. Age year 4	Pop. Age year 3*Exp(-(Z))
		Pop. Age year 5	Pop. Age year 4*Exp(-(Z))
		Pop. Age year 6	Pop. Age year 5*Exp(-(Z))
		Total Sm stock	Sum(Pop.Age yr1:yr6)
34	**Year 2036**		
		Pop. Age year 1	(R B&H a Total Sm Yr34)*Exp(R B&H b*Total Sm Yr34)
		Pop. Age year 2	Pop. Age year 1*Exp(-M)
		Pop. Age year 3	Pop. Age year 2*Exp(-M)
		Pop. Age year 4	Pop. Age year 3*Exp(-(Z))
		Pop. Age year 5	Pop. Age year 4*Exp(-(Z))
		Pop. Age year 6	Pop. Age year 5*Exp(-(Z))
		Total Sm stock	Sum(Pop.Age yr1:yr6)
35	**Year 2037**		
		Pop. Age year 1	(R B&H a Total Sm Yr35)*Exp(R B&H b*Total Sm Yr35)
		Pop. Age year 2	Pop. Age year 1*Exp(-M)
		Pop. Age year 3	Pop. Age year 2*Exp(-M)
		Pop. Age year 4	Pop. Age year 3*Exp(-(Z))
		Pop. Age year 5	Pop. Age year 4*Exp(-(Z))
		Pop. Age year 6	Pop. Age year 5*Exp(-(Z))
		Total Sm stock	Sum(Pop.Age yr1:yr6)

36	**Year 2038**		
		Pop. Age year 1	(R B&H a Total Sm Yr36)*Exp(R B&H b*Total Sm Yr36)
		Pop. Age year 2	Pop. Age year 1*Exp(-M)
		Pop. Age year 3	Pop. Age year 2*Exp(-M)
		Pop. Age year 4	Pop. Age year 3*Exp(-(Z))
		Pop. Age year 5	Pop. Age year 4*Exp(-(Z))
		Pop. Age year 6	Pop. Age year 5*Exp(-(Z))
		Total Sm stock	Sum(Pop.Age yr1:yr6)
37	**Year 2039**		
		Pop. Age year 1	(R B&H a Total Sm Yr37)*Exp(R B&H b*Total Sm Yr37)
		Pop. Age year 2	Pop. Age year 1*Exp(-M)
		Pop. Age year 3	Pop. Age year 2*Exp(-M)
		Pop. Age year 4	Pop. Age year 3*Exp(-(Z))
		Pop. Age year 5	Pop. Age year 4*Exp(-(Z))
		Pop. Age year 6	Pop. Age year 5*Exp(-(Z))
		Total Sm stock	Sum(Pop.Age yr1:yr6)
38	**Year 2040**		
		Pop. Age year 1	(R B&H a Total Sm Yr38)*Exp(R B&H b*Total Sm Yr38)
		Pop. Age year 2	Pop. Age year 1*Exp(-M)
		Pop. Age year 3	Pop. Age year 2*Exp(-M)
		Pop. Age year 4	Pop. Age year 3*Exp(-(Z))
		Pop. Age year 5	Pop. Age year 4*Exp(-(Z))
		Pop. Age year 6	Pop. Age year 5*Exp(-(Z))
		Total Sm stock	Sum(Pop.Age yr1:yr6)
39	**Year 2041**		
		Pop. Age year 1	(R B&H a Total Sm Yr39)*Exp(R B&H b*Total Sm Yr39)
		Pop. Age year 2	Pop. Age year 1*Exp(-M)
		Pop. Age year 3	Pop. Age year 2*Exp(-M)
		Pop. Age year 4	Pop. Age year 3*Exp(-(Z))
		Pop. Age year 5	Pop. Age year 4*Exp(-(Z))
		Pop. Age year 6	Pop. Age year 5*Exp(-(Z))
		Total Sm stock	Sum(Pop.Age yr1:yr6)
40	**Year 2042**		
		Pop. Age year 1	(R B&H a Total Sm Yr40)*Exp(R B&H b*Total Sm Yr40)
		Pop. Age year 2	Pop. Age year 1*Exp(-M)
		Pop. Age year 3	Pop. Age year 2*Exp(-M)
		Pop. Age year 4	Pop. Age year 3*Exp(-(Z))
		Pop. Age year 5	Pop. Age year 4*Exp(-(Z))
		Pop. Age year 6	Pop. Age year 5*Exp(-(Z))
		Total Sm stock	Sum(Pop.Age yr1:yr6)
41	**Year 2043**		
		Pop. Age year 1	(R B&H a Total Sm Yr41)*Exp(R B&H b*Total Sm Yr41)
		Pop. Age year 2	Pop. Age year 1*Exp(-M)
		Pop. Age year 3	Pop. Age year 2*Exp(-M)
		Pop. Age year 4	Pop. Age year 3*Exp(-(Z))
		Pop. Age year 5	Pop. Age year 4*Exp(-(Z))
		Pop. Age year 6	Pop. Age year 5*Exp(-(Z))
		Total Sm stock	Sum(Pop.Age yr1:yr6)
42	**Year 2044**		
		Pop. Age year 1	(R B&H a Total Sm Yr42)*Exp(R B&H b*Total Sm Yr42)
		Pop. Age year 2	Pop. Age year 1*Exp(-M)
		Pop. Age year 3	Pop. Age year 2*Exp(-M)

HAMID AWONG FISHERIES MODEL (HAFM)

		Pop. Age year 4	Pop. Age year 3*Exp(-(Z))
		Pop. Age year 5	Pop. Age year 4*Exp(-(Z))
		Pop. Age year 6	Pop. Age year 5*Exp(-(Z))
		Total Sm stock	Sum(Pop.Age yr1:yr6)

43   **Year 2045**
   Pop. Age year 1   (R B&H a Total Sm Yr43)*Exp(R B&H b*Total Sm Yr43)
   Pop. Age year 2   Pop. Age year 1*Exp(-M)
   Pop. Age year 3   Pop. Age year 2*Exp(-M)
   Pop. Age year 4   Pop. Age year 3*Exp(-(Z))
   Pop. Age year 5   Pop. Age year 4*Exp(-(Z))
   Pop. Age year 6   Pop. Age year 5*Exp(-(Z))
   Total Sm stock   Sum(Pop.Age yr1:yr6)

44   **Year 2046**
   Pop. Age year 1   (R B&H a Total Sm Yr44)*Exp(R B&H b*Total Sm Yr44)
   Pop. Age year 2   Pop. Age year 1*Exp(-M)
   Pop. Age year 3   Pop. Age year 2*Exp(-M)
   Pop. Age year 4   Pop. Age year 3*Exp(-(Z))
   Pop. Age year 5   Pop. Age year 4*Exp(-(Z))
   Pop. Age year 6   Pop. Age year 5*Exp(-(Z))
   Total Sm stock   Sum(Pop.Age yr1:yr6)

45   **Year 2047**
   Pop. Age year 1   (R B&H a Total Sm Yr45)*Exp(R B&H b*Total Sm Yr45)
   Pop. Age year 2   Pop. Age year 1*Exp(-M)
   Pop. Age year 3   Pop. Age year 2*Exp(-M)
   Pop. Age year 4   Pop. Age year 3*Exp(-(Z))
   Pop. Age year 5   Pop. Age year 4*Exp(-(Z))
   Pop. Age year 6   Pop. Age year 5*Exp(-(Z))
   Total Sm stock   Sum(Pop.Age yr1:yr6)

46   **Year 2048**
   Pop. Age year 1   (R B&H a Total Sm Yr46)*Exp(R B&H b*Total Sm Yr46)
   Pop. Age year 2   Pop. Age year 1*Exp(-M)
   Pop. Age year 3   Pop. Age year 2*Exp(-M)
   Pop. Age year 4   Pop. Age year 3*Exp(-(Z))
   Pop. Age year 5   Pop. Age year 4*Exp(-(Z))
   Pop. Age year 6   Pop. Age year 5*Exp(-(Z))
   Total Sm stock   Sum(Pop.Age yr1:yr6)

47   **Year 2049**
   Pop. Age year 1   (R B&H a Total Sm Yr47)*Exp(R B&H b*Total Sm Yr47)
   Pop. Age year 2   Pop. Age year 1*Exp(-M)
   Pop. Age year 3   Pop. Age year 2*Exp(-M)
   Pop. Age year 4   Pop. Age year 3*Exp(-(Z))
   Pop. Age year 5   Pop. Age year 4*Exp(-(Z))
   Pop. Age year 6   Pop. Age year 5*Exp(-(Z))
   Total Sm stock   Sum(Pop.Age yr1:yr6)

48   **Year 2050**
   Pop. Age year 1   (R B&H a Total Sm Yr48)*Exp(R B&H b*Total Sm Yr48)
   Pop. Age year 2   Pop. Age year 1*Exp(-M)
   Pop. Age year 3   Pop. Age year 2*Exp(-M)
   Pop. Age year 4   Pop. Age year 3*Exp(-(Z))
   Pop. Age year 5   Pop. Age year 4*Exp(-(Z))
   Pop. Age year 6   Pop. Age year 5*Exp(-(Z))
   Total Sm stock   Sum(Pop.Age yr1:yr6)

49	Year 2051		
		Pop. Age year 1	(R B&H a Total Sm Yr49)*Exp(R B&H b*Total Sm Yr49)
		Pop. Age year 2	Pop. Age year 1*Exp(-M)
		Pop. Age year 3	Pop. Age year 2*Exp(-M)
		Pop. Age year 4	Pop. Age year 3*Exp(-(Z))
		Pop. Age year 5	Pop. Age year 4*Exp(-(Z))
		Pop. Age year 6	Pop. Age year 5*Exp(-(Z))
		Total Sm stock	Sum(Pop.Age yr1:yr6)
50	Year 2052		
		Pop. Age year 1	(R B&H a Total Sm Yr50)*Exp(R B&H b*Total Sm Yr50)
		Pop. Age year 2	Pop. Age year 1*Exp(-M)
		Pop. Age year 3	Pop. Age year 2*Exp(-M)
		Pop. Age year 4	Pop. Age year 3*Exp(-(Z))
		Pop. Age year 5	Pop. Age year 4*Exp(-(Z))
		Pop. Age year 6	Pop. Age year 5*Exp(-(Z))
		Total Sm stock	Sum(Pop.Age yr1:yr6)

6.3.3 Biomass Model (Yield)

Assumed the initials yield of stock in 2003 was 158.2663257 gram after six modes through natural mortality and fishing mortality the stock biomass was 755.0916870 gram. In 50 years later or 2052 the biomass stock through recruitment, natural mortality (M) and Fishing mortality (F) the total biomass for the six modes would be 4.0×10^{-7} gram as shown in Table 6.6 and Figure 6.3 The fallen of biomass stock was due to the low recruitment, natural mortality and fishing mortality.

Table 6.6 The Biomass (yield) of *Priacanthus tayenus* for 50 years from 2003 to 2052

Mean Wt	33.3982528	104.671604	178.912996	237.917315	279.252353	6306.358785	
Year	1	2	3	4	5	6	Total
2003	158.2663257	226.4679655	176.7392115	107.3075308	57.5061269	28.8045265	755.0916870
2004	32.4042222	226.4679655	176.7392115	103.2030938	55.3065584	27.7027738	621.8238253
2005	31.9114034	46.36815985	176.7392115	103.2030938	53.1911217	26.6431624	438.0561527
2006	31.7628204	45.6629708	36.186451338	103.2030938	53.1911217	25.6240803	295.6305384
2007	16.3201153	45.4503589	35.6361106	21.1303066	53.1911217	25.6240803	197.3520934
2008	9.4569468	23.3529355	35.470184	20.8089469	10.8906106	25.6240803	125.6037050
2009	6.4207216	13.5322249	18.2250033	20.7120581	10.7249811	5.2463996	74.8613885
2010	4.4962319	9.1876004	10.5607641	10.6421020	10.6750443	5.1666099	50.7283527
2011	2.8089428	6.4337912	7.1701499	6.1667330	5.4849649	5.1425536	33.2071353
2012	1.6944648	4.0193993	5.0210332	4.1868561	3.1783490	2.6423053	20.7424076

HAMID AWONG FISHERIES MODEL (HAFM)

Year							
2013	1.1287844	2.4246598	3.1368033	2.9319253	2.1579157	1.5311252	13.3112137
2014	0.7437418	1.6152110	1.8922431	1.8316694	1.5111213	1.0395457	8.6335324
2015	0.4687517	1.0642421	1.2605364	1.1049350	0.9440468	0.7279616	5.5704737
2016	0.2982479	0.6707507	0.8305515	0.7360634	0.5694862	0.4547814	3.5598810
2017	0.1934600	0.4267717	0.5234645	0.4849829	0.3793688	0.2743420	2.2823899
2018	0.1251331	0.2768276	0.3330594	0.3056660	0.2499613	0.1827556	1.4734030
2019	0.0799840	0.1790566	0.2160407	0.1944830	0.1575410	0.1204154	0.9475206
2020	0.0512412	0.1144515	0.1397387	0.1261523	0.1002370	0.0758932	0.6077139
2021	0.0330556	0.0733226	0.0893198	0.0815974	0.0650192	0.0482878	0.3906024
2022	0.0212762	0.0473003	0.0572221	0.0521564	0.0420555	0.0313221	0.2513326
2023	0.0136490	0.0304447	0.0369139	0.0334136	0.0268815	0.0202597	0.1615625
2024	0.0087680	0.0195308	0.0237595	0.0215551	0.0172215	0.0129498	0.1037848
2025	0.0056418	0.0125465	0.0152422	0.0138739	0.0111095	0.0082962	0.0667101
2026	0.0036275	0.0080731	0.0097915	0.0089003	0.0071506	0.0053519	0.0428949
2027	0.0023303	0.0051907	0.0063004	0.0057175	0.0045873	0.0034447	0.0275709
2028	0.0014977	0.0033345	0.0040509	0.0036790	0.0029468	0.0022098	0.0177187
2029	0.0009630	0.0021430	0.0026023	0.0023655	0.0018961	0.0014196	0.0113895
2030	0.0006190	0.0013780	0.0016725	0.0015195	0.0012192	0.0009134	0.0073216
2031	0.0003978	0.0008858	0.0010754	0.0009766	0.0007832	0.0005873	0.0047061
2032	0.0002557	0.0005692	0.0006913	0.0006280	0.0005033	0.0003773	0.0030248
2033	0.0001644	0.0003659	0.0004442	0.0004037	0.0003236	0.0002425	0.0019443
2034	0.0001057	0.0002352	0.0002855	0.0002594	0.0002080	0.0001559	0.0012498
2035	0.0000679	0.0001512	0.0001836	0.0001667	0.0001337	0.0001002	0.0008033
2036	0.0000437	0.0000972	0.0001180	0.0001072	0.0000859	0.0000644	0.0005164
2037	0.0000281	0.0000625	0.0000758	0.0000689	0.0000552	0.0000414	0.0003319
2038	0.0000180	0.0000402	0.0000487	0.0000443	0.0000355	0.0000266	0.0002133
2039	0.0000116	0.0000258	0.0000313	0.0000285	0.0000228	0.0000171	0.0001371
2040	0.0000075	0.0000166	0.0000201	0.0000183	0.0000147	0.0000110	0.0000881
2041	0.0000048	0.0000107	0.0000129	0.0000118	0.0000094	0.0000071	0.0000567
2042	0.0000031	0.0000069	0.0000083	0.0000076	0.0000061	0.0000045	0.0000364
2043	0.0000020	0.0000044	0.0000053	0.0000049	0.0000039	0.0000029	0.0000234
2044	0.0000013	0.0000028	0.0000034	0.0000031	0.0000025	0.0000019	0.0000150
2045	0.0000008	0.0000018	0.0000022	0.0000020	0.0000016	0.0000012	0.0000097
2046	0.0000005	0.0000012	0.0000014	0.0000013	0.0000010	0.0000008	0.0000062
2047	0.0000003	0.0000008	0.0000009	0.0000008	0.0000007	0.0000005	0.0000040
2048	0.0000002	0.0000005	0.0000006	0.0000005	0.0000004	0.0000003	0.0000026
2049	0.0000001	0.0000003	0.0000004	0.0000003	0.0000003	0.0000002	0.0000017
2050	0.0000001	0.0000002	0.0000002	0.0000002	0.0000002	0.0000001	0.0000011
2051	0.0000001	0.0000001	0.0000002	0.0000001	0.0000001	0.0000001	0.0000007
2052	0.0000000	0.0000001	0.0000001	0.0000001	0.0000001	0.0000001	0.0000004

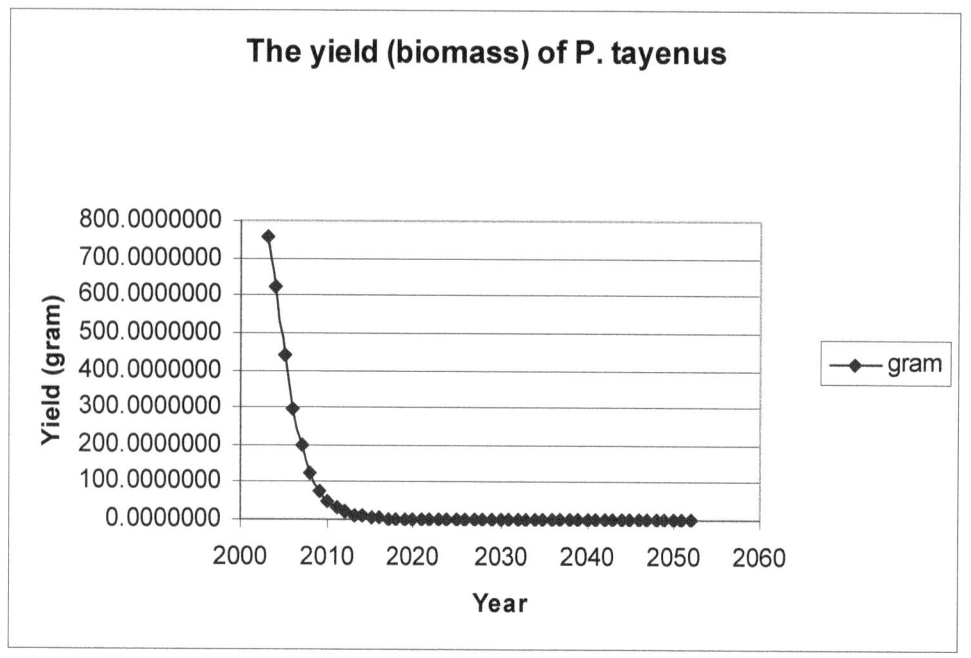

Figure 6.6 The Biomass (Yield) of *Priacanthus tayenus* for 50 Years (2003 To 2052)

Design Model the Yield (Biomass) of the Stock

Mean Weight

Age year 1	Winf*(1-Exp(-K*(yr1+to)))^b
Age year 2	Winf*(1-Exp(-K*(yr1+to)))^b
Age year 3	Winf*(1-Exp(-K*(yr1+to)))^b
Age year 4	Winf*(1-Exp(-K*(yr1+to)))^b
Age year 5	Winf*(1-Exp(-K*(yr1+to)))^b
Age year 6	Winf*(1-Exp(-K*(yr1+to)))^b

          Year 2003

1

Yield Age year 1	Pop. Yr1*(F/Z)*mean wt yr1
Yield Age year 2	Pop. Yr2*(F/Z)*mean wt yr2
Yield Age year 3	Pop. Yr3*(F/Z)*mean wt yr3
Yield Age year 4	Pop. Yr4*(F/Z)*mean wt yr4
Yield Age year 5	Pop. Yr5*(F/Z)*mean wt yr5
Yield Age year 6	Pop. Yr6*(F/Z)*mean wt yr6
Total Yield	Sum(Yield.Age yr1:yr6)

2   **Year 2004**

Yield Age year 1	Pop. Yr1*(F/Z)*mean wt yr1
Yield Age year 2	Pop. Yr2*(F/Z)*mean wt yr2
Yield Age year 3	Pop. Yr3*(F/Z)*mean wt yr3
Yield Age year 4	Pop. Yr4*(F/Z)*mean wt yr4
Yield Age year 5	Pop. Yr5*(F/Z)*mean wt yr5
Yield Age year 6	Pop. Yr6*(F/Z)*mean wt yr6
Total Yield	Sum(Yield.Age yr1:yr6)

3   **Year 2005**

		Yield Age year 1	Pop. Yr1*(F/Z)*mean wt yr1
		Yield Age year 2	Pop. Yr2*(F/Z)*mean wt yr2
		Yield Age year 3	Pop. Yr3*(F/Z)*mean wt yr3
		Yield Age year 4	Pop. Yr4*(F/Z)*mean wt yr4
		Yield Age year 5	Pop. Yr5*(F/Z)*mean wt yr5
		Yield Age year 6	Pop. Yr6*(F/Z)*mean wt yr6
		Total Yield	Sum(Yield.Age yr1:yr6)
4	**Year 2006**		
		Yield Age year 1	Pop. Yr1*(F/Z)*mean wt yr1
		Yield Age year 2	Pop. Yr2*(F/Z)*mean wt yr2
		Yield Age year 3	Pop. Yr3*(F/Z)*mean wt yr3
		Yield Age year 4	Pop. Yr4*(F/Z)*mean wt yr4
		Yield Age year 5	Pop. Yr5*(F/Z)*mean wt yr5
		Yield Age year 6	Pop. Yr6*(F/Z)*mean wt yr6
		Total Yield	Sum(Yield.Age yr1:yr6)
5	**Year 2007**		
		Yield Age year 1	Pop. Yr1*(F/Z)*mean wt yr1
		Yield Age year 2	Pop. Yr2*(F/Z)*mean wt yr2
		Yield Age year 3	Pop. Yr3*(F/Z)*mean wt yr3
		Yield Age year 4	Pop. Yr4*(F/Z)*mean wt yr4
		Yield Age year 5	Pop. Yr5*(F/Z)*mean wt yr5
		Yield Age year 6	Pop. Yr6*(F/Z)*mean wt yr6
		Total Yield	Sum(Yield.Age yr1:yr6)
6	**Year 2008**		
		Yield Age year 1	Pop. Yr1*(F/Z)*mean wt yr1
		Yield Age year 2	Pop. Yr2*(F/Z)*mean wt yr2
		Yield Age year 3	Pop. Yr3*(F/Z)*mean wt yr3
		Yield Age year 4	Pop. Yr4*(F/Z)*mean wt yr4
		Yield Age year 5	Pop. Yr5*(F/Z)*mean wt yr5
		Yield Age year 6	Pop. Yr6*(F/Z)*mean wt yr6
		Total Yield	Sum(Yield.Age yr1:yr6)
7	**Year 2009**		
		Yield Age year 1	Pop. Yr1*(F/Z)*mean wt yr1
		Yield Age year 2	Pop. Yr2*(F/Z)*mean wt yr2
		Yield Age year 3	Pop. Yr3*(F/Z)*mean wt yr3
		Yield Age year 4	Pop. Yr4*(F/Z)*mean wt yr4
		Yield Age year 5	Pop. Yr5*(F/Z)*mean wt yr5
		Yield Age year 6	Pop. Yr6*(F/Z)*mean wt yr6
		Total Yield	Sum(Yield.Age yr1:yr6)
8	**Year 2010**		
		Yield Age year 1	Pop. Yr1*(F/Z)*mean wt yr1
		Yield Age year 2	Pop. Yr2*(F/Z)*mean wt yr2
		Yield Age year 3	Pop. Yr3*(F/Z)*mean wt yr3
		Yield Age year 4	Pop. Yr4*(F/Z)*mean wt yr4
		Yield Age year 5	Pop. Yr5*(F/Z)*mean wt yr5
		Yield Age year 6	Pop. Yr6*(F/Z)*mean wt yr6
		Total Yield	Sum(Yield.Age yr1:yr6)
9	**Year 2011**		
		Yield Age year 1	Pop. Yr1*(F/Z)*mean wt yr1
		Yield Age year 2	Pop. Yr2*(F/Z)*mean wt yr2
		Yield Age year 3	Pop. Yr3*(F/Z)*mean wt yr3
		Yield Age year 4	Pop. Yr4*(F/Z)*mean wt yr4
		Yield Age year 5	Pop. Yr5*(F/Z)*mean wt yr5

		Yield Age year 6	Pop. Yr6*(F/Z)*mean wt yr6
		Total Yield	Sum(Yield.Age yr1:yr6)
10	Year 2012		
		Yield Age year 1	Pop. Yr1*(F/Z)*mean wt yr1
		Yield Age year 2	Pop. Yr2*(F/Z)*mean wt yr2
		Yield Age year 3	Pop. Yr3*(F/Z)*mean wt yr3
		Yield Age year 4	Pop. Yr4*(F/Z)*mean wt yr4
		Yield Age year 5	Pop. Yr5*(F/Z)*mean wt yr5
		Yield Age year 6	Pop. Yr6*(F/Z)*mean wt yr6
		Total Yield	Sum(Yield.Age yr1:yr6)
11	Year 2013		
		Yield Age year 1	Pop. Yr1*(F/Z)*mean wt yr1
		Yield Age year 2	Pop. Yr2*(F/Z)*mean wt yr2
		Yield Age year 3	Pop. Yr3*(F/Z)*mean wt yr3
		Yield Age year 4	Pop. Yr4*(F/Z)*mean wt yr4
		Yield Age year 5	Pop. Yr5*(F/Z)*mean wt yr5
		Yield Age year 6	Pop. Yr6*(F/Z)*mean wt yr6
		Total Yield	Sum(Yield.Age yr1:yr6)
12	Year 2014		
		Yield Age year 1	Pop. Yr1*(F/Z)*mean wt yr1
		Yield Age year 2	Pop. Yr2*(F/Z)*mean wt yr2
		Yield Age year 3	Pop. Yr3*(F/Z)*mean wt yr3
		Yield Age year 4	Pop. Yr4*(F/Z)*mean wt yr4
		Yield Age year 5	Pop. Yr5*(F/Z)*mean wt yr5
		Yield Age year 6	Pop. Yr6*(F/Z)*mean wt yr6
		Total Yield	Sum(Yield.Age yr1:yr6)
13	Year 2015		
		Yield Age year 1	Pop. Yr1*(F/Z)*mean wt yr1
		Yield Age year 2	Pop. Yr2*(F/Z)*mean wt yr2
		Yield Age year 3	Pop. Yr3*(F/Z)*mean wt yr3
		Yield Age year 4	Pop. Yr4*(F/Z)*mean wt yr4
		Yield Age year 5	Pop. Yr5*(F/Z)*mean wt yr5
		Yield Age year 6	Pop. Yr6*(F/Z)*mean wt yr6
		Total Yield	Sum(Yield.Age yr1:yr6)
14	Year 2016		
		Yield Age year 1	Pop. Yr1*(F/Z)*mean wt yr1
		Yield Age year 2	Pop. Yr2*(F/Z)*mean wt yr2
		Yield Age year 3	Pop. Yr3*(F/Z)*mean wt yr3
		Yield Age year 4	Pop. Yr4*(F/Z)*mean wt yr4
		Yield Age year 5	Pop. Yr5*(F/Z)*mean wt yr5
		Yield Age year 6	Pop. Yr6*(F/Z)*mean wt yr6
		Total Yield	Sum(Yield.Age yr1:yr6)
15	Year 2017		
		Yield Age year 1	Pop. Yr1*(F/Z)*mean wt yr1
		Yield Age year 2	Pop. Yr2*(F/Z)*mean wt yr2
		Yield Age year 3	Pop. Yr3*(F/Z)*mean wt yr3
		Yield Age year 4	Pop. Yr4*(F/Z)*mean wt yr4
		Yield Age year 5	Pop. Yr5*(F/Z)*mean wt yr5
		Yield Age year 6	Pop. Yr6*(F/Z)*mean wt yr6
		Total Yield	Sum(Yield.Age yr1:yr6)
16	Year 2018		
		Yield Age year 1	Pop. Yr1*(F/Z)*mean wt yr1

		Yield Age year 2	Pop. Yr2*(F/Z)*mean wt yr2
		Yield Age year 3	Pop. Yr3*(F/Z)*mean wt yr3
		Yield Age year 4	Pop. Yr4*(F/Z)*mean wt yr4
		Yield Age year 5	Pop. Yr5*(F/Z)*mean wt yr5
		Yield Age year 6	Pop. Yr6*(F/Z)*mean wt yr6
		Total Yield	Sum(Yield.Age yr1:yr6)
17	**Year 2019**		
		Yield Age year 1	Pop. Yr1*(F/Z)*mean wt yr1
		Yield Age year 2	Pop. Yr2*(F/Z)*mean wt yr2
		Yield Age year 3	Pop. Yr3*(F/Z)*mean wt yr3
		Yield Age year 4	Pop. Yr4*(F/Z)*mean wt yr4
		Yield Age year 5	Pop. Yr5*(F/Z)*mean wt yr5
		Yield Age year 6	Pop. Yr6*(F/Z)*mean wt yr6
		Total Yield	Sum(Yield.Age yr1:yr6)
18	**Year 2020**		
		Yield Age year 1	Pop. Yr1*(F/Z)*mean wt yr1
		Yield Age year 2	Pop. Yr2*(F/Z)*mean wt yr2
		Yield Age year 3	Pop. Yr3*(F/Z)*mean wt yr3
		Yield Age year 4	Pop. Yr4*(F/Z)*mean wt yr4
		Yield Age year 5	Pop. Yr5*(F/Z)*mean wt yr5
		Yield Age year 6	Pop. Yr6*(F/Z)*mean wt yr6
		Total Yield	Sum(Yield.Age yr1:yr6)
19	**Year 2021**		
		Yield Age year 1	Pop. Yr1*(F/Z)*mean wt yr1
		Yield Age year 2	Pop. Yr2*(F/Z)*mean wt yr2
		Yield Age year 3	Pop. Yr3*(F/Z)*mean wt yr3
		Yield Age year 4	Pop. Yr4*(F/Z)*mean wt yr4
		Yield Age year 5	Pop. Yr5*(F/Z)*mean wt yr5
		Yield Age year 6	Pop. Yr6*(F/Z)*mean wt yr6
		Total Yield	Sum(Yield.Age yr1:yr6)
20	**Year 2022**		
		Pop. Age year 1	Pop. Yr1*(F/Z)*mean wt yr1
		Pop. Age year 2	Pop. Yr2*(F/Z)*mean wt yr2
		Pop. Age year 3	Pop. Yr3*(F/Z)*mean wt yr3
		Pop. Age year 4	Pop. Yr4*(F/Z)*mean wt yr4
		Pop. Age year 5	Pop. Yr5*(F/Z)*mean wt yr5
		Pop. Age year 6	Pop. Yr6*(F/Z)*mean wt yr6
		Total Yield	Sum(Yield.Age yr1:yr6)
21	**Year 2023**		
		Pop. Age year 1	Pop. Yr1*(F/Z)*mean wt yr1
		Pop. Age year 2	Pop. Yr2*(F/Z)*mean wt yr2
		Pop. Age year 3	Pop. Yr3*(F/Z)*mean wt yr3
		Pop. Age year 4	Pop. Yr4*(F/Z)*mean wt yr4
		Pop. Age year 5	Pop. Yr5*(F/Z)*mean wt yr5
		Pop. Age year 6	Pop. Yr6*(F/Z)*mean wt yr6
		Total Yield	Sum(Yield.Age yr1:yr6)
22	**Year 2024**		
		Pop. Age year 1	Pop. Yr1*(F/Z)*mean wt yr1
		Pop. Age year 2	Pop. Yr2*(F/Z)*mean wt yr2
		Pop. Age year 3	Pop. Yr3*(F/Z)*mean wt yr3
		Pop. Age year 4	Pop. Yr4*(F/Z)*mean wt yr4
		Pop. Age year 5	Pop. Yr5*(F/Z)*mean wt yr5
		Pop. Age year 6	Pop. Yr6*(F/Z)*mean wt yr6

			Total Yield	Sum(Yield.Age yr1:yr6)
23	**Year 2025**			
			Pop. Age year 1	Pop. Yr1*(F/Z)*mean wt yr1
			Pop. Age year 2	Pop. Yr2*(F/Z)*mean wt yr2
			Pop. Age year 3	Pop. Yr3*(F/Z)*mean wt yr3
			Pop. Age year 4	Pop. Yr4*(F/Z)*mean wt yr4
			Pop. Age year 5	Pop. Yr5*(F/Z)*mean wt yr5
			Pop. Age year 6	Pop. Yr6*(F/Z)*mean wt yr6
			Total Yield	Sum(Yield.Age yr1:yr6)
24	**Year 2026**			
			Pop. Age year 1	Pop. Yr1*(F/Z)*mean wt yr1
			Pop. Age year 2	Pop. Yr2*(F/Z)*mean wt yr2
			Pop. Age year 3	Pop. Yr3*(F/Z)*mean wt yr3
			Pop. Age year 4	Pop. Yr4*(F/Z)*mean wt yr4
			Pop. Age year 5	Pop. Yr5*(F/Z)*mean wt yr5
			Pop. Age year 6	Pop. Yr6*(F/Z)*mean wt yr6
			Total Yield	Sum(Yield.Age yr1:yr6)
25	**Year 2027**			
			Pop. Age year 1	Pop. Yr1*(F/Z)*mean wt yr1
			Pop. Age year 2	Pop. Yr2*(F/Z)*mean wt yr2
			Pop. Age year 3	Pop. Yr3*(F/Z)*mean wt yr3
			Pop. Age year 4	Pop. Yr4*(F/Z)*mean wt yr4
			Pop. Age year 5	Pop. Yr5*(F/Z)*mean wt yr5
			Pop. Age year 6	Pop. Yr6*(F/Z)*mean wt yr6
			Total Yield	Sum(Yield.Age yr1:yr6)
26	**Year 2028**			
			Pop. Age year 1	Pop. Yr1*(F/Z)*mean wt yr1
			Pop. Age year 2	Pop. Yr2*(F/Z)*mean wt yr2
			Pop. Age year 3	Pop. Yr3*(F/Z)*mean wt yr3
			Pop. Age year 4	Pop. Yr4*(F/Z)*mean wt yr4
			Pop. Age year 5	Pop. Yr5*(F/Z)*mean wt yr5
			Pop. Age year 6	Pop. Yr6*(F/Z)*mean wt yr6
			Total Yield	Sum(Yield.Age yr1:yr6)
27	**Year 2029**			
			Pop. Age year 1	Pop. Yr1*(F/Z)*mean wt yr1
			Pop. Age year 2	Pop. Yr2*(F/Z)*mean wt yr2
			Pop. Age year 3	Pop. Yr3*(F/Z)*mean wt yr3
			Pop. Age year 4	Pop. Yr4*(F/Z)*mean wt yr4
			Pop. Age year 5	Pop. Yr5*(F/Z)*mean wt yr5
			Pop. Age year 6	Pop. Yr6*(F/Z)*mean wt yr6
			Total Yield	Sum(Yield.Age yr1:yr6)
28	**Year 2030**			
			Pop. Age year 1	Pop. Yr1*(F/Z)*mean wt yr1
			Pop. Age year 2	Pop. Yr2*(F/Z)*mean wt yr2
			Pop. Age year 3	Pop. Yr3*(F/Z)*mean wt yr3
			Pop. Age year 4	Pop. Yr4*(F/Z)*mean wt yr4
			Pop. Age year 5	Pop. Yr5*(F/Z)*mean wt yr5
			Pop. Age year 6	Pop. Yr6*(F/Z)*mean wt yr6
			Total Yield	Sum(Yield.Age yr1:yr6)
29	**Year 2031**			
			Pop. Age year 1	Pop. Yr1*(F/Z)*mean wt yr1
			Pop. Age year 2	Pop. Yr2*(F/Z)*mean wt yr2

HAMID AWONG FISHERIES MODEL (HAFM)

		Pop. Age year 3	Pop. Yr3*(F/Z)*mean wt yr3
		Pop. Age year 4	Pop. Yr4*(F/Z)*mean wt yr4
		Pop. Age year 5	Pop. Yr5*(F/Z)*mean wt yr5
		Pop. Age year 6	Pop. Yr6*(F/Z)*mean wt yr6
		Total Yield	Sum(Yield.Age yr1:yr6)
30	**Year 2032**		
		Pop. Age year 1	Pop. Yr1*(F/Z)*mean wt yr1
		Pop. Age year 2	Pop. Yr2*(F/Z)*mean wt yr2
		Pop. Age year 3	Pop. Yr3*(F/Z)*mean wt yr3
		Pop. Age year 4	Pop. Yr4*(F/Z)*mean wt yr4
		Pop. Age year 5	Pop. Yr5*(F/Z)*mean wt yr5
		Pop. Age year 6	Pop. Yr6*(F/Z)*mean wt yr6
		Total Yield	Sum(Yield.Age yr1:yr6)
31	**Year 2033**		
		Pop. Age year 1	Pop. Yr1*(F/Z)*mean wt yr1
		Pop. Age year 2	Pop. Yr2*(F/Z)*mean wt yr2
		Pop. Age year 3	Pop. Yr3*(F/Z)*mean wt yr3
		Pop. Age year 4	Pop. Yr4*(F/Z)*mean wt yr4
		Pop. Age year 5	Pop. Yr5*(F/Z)*mean wt yr5
		Pop. Age year 6	Pop. Yr6*(F/Z)*mean wt yr6
		Total Yield	Sum(Yield.Age yr1:yr6)
32	**Year 2034**		
		Pop. Age year 1	Pop. Yr1*(F/Z)*mean wt yr1
		Pop. Age year 2	Pop. Yr2*(F/Z)*mean wt yr2
		Pop. Age year 3	Pop. Yr3*(F/Z)*mean wt yr3
		Pop. Age year 4	Pop. Yr4*(F/Z)*mean wt yr4
		Pop. Age year 5	Pop. Yr5*(F/Z)*mean wt yr5
		Pop. Age year 6	Pop. Yr6*(F/Z)*mean wt yr6
		Total Yield	Sum(Yield.Age yr1:yr6)
33	**Year 2035**		
		Pop. Age year 1	Pop. Yr1*(F/Z)*mean wt yr1
		Pop. Age year 2	Pop. Yr2*(F/Z)*mean wt yr2
		Pop. Age year 3	Pop. Yr3*(F/Z)*mean wt yr3
		Pop. Age year 4	Pop. Yr4*(F/Z)*mean wt yr4
		Pop. Age year 5	Pop. Yr5*(F/Z)*mean wt yr5
		Pop. Age year 6	Pop. Yr6*(F/Z)*mean wt yr6
		Total Yield	Sum(Yield.Age yr1:yr6)
34	**Year 2036**		
		Pop. Age year 1	Pop. Yr1*(F/Z)*mean wt yr1
		Pop. Age year 2	Pop. Yr2*(F/Z)*mean wt yr2
		Pop. Age year 3	Pop. Yr3*(F/Z)*mean wt yr3
		Pop. Age year 4	Pop. Yr4*(F/Z)*mean wt yr4
		Pop. Age year 5	Pop. Yr5*(F/Z)*mean wt yr5
		Pop. Age year 6	Pop. Yr6*(F/Z)*mean wt yr6
		Total Yield	Sum(Yield.Age yr1:yr6)
35	**Year 2037**		
		Pop. Age year 1	Pop. Yr1*(F/Z)*mean wt yr1
		Pop. Age year 2	Pop. Yr2*(F/Z)*mean wt yr2
		Pop. Age year 3	Pop. Yr3*(F/Z)*mean wt yr3
		Pop. Age year 4	Pop. Yr4*(F/Z)*mean wt yr4
		Pop. Age year 5	Pop. Yr5*(F/Z)*mean wt yr5
		Pop. Age year 6	Pop. Yr6*(F/Z)*mean wt yr6
		Total Yield	Sum(Yield.Age yr1:yr6)

36	**Year 2038**		
		Pop. Age year 1	Pop. Yr1*(F/Z)*mean wt yr1
		Pop. Age year 2	Pop. Yr2*(F/Z)*mean wt yr2
		Pop. Age year 3	Pop. Yr3*(F/Z)*mean wt yr3
		Pop. Age year 4	Pop. Yr4*(F/Z)*mean wt yr4
		Pop. Age year 5	Pop. Yr5*(F/Z)*mean wt yr5
		Pop. Age year 6	Pop. Yr6*(F/Z)*mean wt yr6
		Total Yield	Sum(Yield.Age yr1:yr6)
37	**Year 2039**		
		Pop. Age year 1	Pop. Yr1*(F/Z)*mean wt yr1
		Pop. Age year 2	Pop. Yr2*(F/Z)*mean wt yr2
		Pop. Age year 3	Pop. Yr3*(F/Z)*mean wt yr3
		Pop. Age year 4	Pop. Yr4*(F/Z)*mean wt yr4
		Pop. Age year 5	Pop. Yr5*(F/Z)*mean wt yr5
		Pop. Age year 6	Pop. Yr6*(F/Z)*mean wt yr6
		Total Yield	Sum(Yield.Age yr1:yr6)
38	**Year 2040**		
		Pop. Age year 1	Pop. Yr1*(F/Z)*mean wt yr1
		Pop. Age year 2	Pop. Yr2*(F/Z)*mean wt yr2
		Pop. Age year 3	Pop. Yr3*(F/Z)*mean wt yr3
		Pop. Age year 4	Pop. Yr4*(F/Z)*mean wt yr4
		Pop. Age year 5	Pop. Yr5*(F/Z)*mean wt yr5
		Pop. Age year 6	Pop. Yr6*(F/Z)*mean wt yr6
		Total Yield	Sum(Yield.Age yr1:yr6)
39	**Year 2041**		
		Pop. Age year 1	Pop. Yr1*(F/Z)*mean wt yr1
		Pop. Age year 2	Pop. Yr2*(F/Z)*mean wt yr2
		Pop. Age year 3	Pop. Yr3*(F/Z)*mean wt yr3
		Pop. Age year 4	Pop. Yr4*(F/Z)*mean wt yr4
		Pop. Age year 5	Pop. Yr5*(F/Z)*mean wt yr5
		Pop. Age year 6	Pop. Yr6*(F/Z)*mean wt yr6
		Total Yield	Sum(Yield.Age yr1:yr6)
40	**Year 2042**		
		Pop. Age year 1	Pop. Yr1*(F/Z)*mean wt yr1
		Pop. Age year 2	Pop. Yr2*(F/Z)*mean wt yr2
		Pop. Age year 3	Pop. Yr3*(F/Z)*mean wt yr3
		Pop. Age year 4	Pop. Yr4*(F/Z)*mean wt yr4
		Pop. Age year 5	Pop. Yr5*(F/Z)*mean wt yr5
		Pop. Age year 6	Pop. Yr6*(F/Z)*mean wt yr6
		Total Yield	Sum(Yield.Age yr1:yr6)
41	**Year 2043**		
		Pop. Age year 1	Pop. Yr1*(F/Z)*mean wt yr1
		Pop. Age year 2	Pop. Yr2*(F/Z)*mean wt yr2
		Pop. Age year 3	Pop. Yr3*(F/Z)*mean wt yr3
		Pop. Age year 4	Pop. Yr4*(F/Z)*mean wt yr4
		Pop. Age year 5	Pop. Yr5*(F/Z)*mean wt yr5
		Pop. Age year 6	Pop. Yr6*(F/Z)*mean wt yr6
		Total Yield	Sum(Yield.Age yr1:yr6)
42	**Year 2044**		
		Pop. Age year 1	Pop. Yr1*(F/Z)*mean wt yr1
		Pop. Age year 2	Pop. Yr2*(F/Z)*mean wt yr2
		Pop. Age year 3	Pop. Yr3*(F/Z)*mean wt yr3

HAMID AWONG FISHERIES MODEL (HAFM)

		Pop. Age year 4	Pop. Yr4*(F/Z)*mean wt yr4
		Pop. Age year 5	Pop. Yr5*(F/Z)*mean wt yr5
		Pop. Age year 6	Pop. Yr6*(F/Z)*mean wt yr6
		Total Yield	Sum(Yield.Age yr1:yr6)
43	**Year 2045**		
		Pop. Age year 1	Pop. Yr1*(F/Z)*mean wt yr1
		Pop. Age year 2	Pop. Yr2*(F/Z)*mean wt yr2
		Pop. Age year 3	Pop. Yr3*(F/Z)*mean wt yr3
		Pop. Age year 4	Pop. Yr4*(F/Z)*mean wt yr4
		Pop. Age year 5	Pop. Yr5*(F/Z)*mean wt yr5
		Pop. Age year 6	Pop. Yr6*(F/Z)*mean wt yr6
		Total Yield	Sum(Yield.Age yr1:yr6)
44	**Year 2046**		
		Pop. Age year 1	Pop. Yr1*(F/Z)*mean wt yr1
		Pop. Age year 2	Pop. Yr2*(F/Z)*mean wt yr2
		Pop. Age year 3	Pop. Yr3*(F/Z)*mean wt yr3
		Pop. Age year 4	Pop. Yr4*(F/Z)*mean wt yr4
		Pop. Age year 5	Pop. Yr5*(F/Z)*mean wt yr5
		Pop. Age year 6	Pop. Yr6*(F/Z)*mean wt yr6
		Total Yield	Sum(Yield.Age yr1:yr6)
45	**Year 2047**		
		Pop. Age year 1	Pop. Yr1*(F/Z)*mean wt yr1
		Pop. Age year 2	Pop. Yr2*(F/Z)*mean wt yr2
		Pop. Age year 3	Pop. Yr3*(F/Z)*mean wt yr3
		Pop. Age year 4	Pop. Yr4*(F/Z)*mean wt yr4
		Pop. Age year 5	Pop. Yr5*(F/Z)*mean wt yr5
		Pop. Age year 6	Pop. Yr6*(F/Z)*mean wt yr6
		Total Yield	Sum(Yield.Age yr1:yr6)
46	**Year 2048**		
		Pop. Age year 1	Pop. Yr1*(F/Z)*mean wt yr1
		Pop. Age year 2	Pop. Yr2*(F/Z)*mean wt yr2
		Pop. Age year 3	Pop. Yr3*(F/Z)*mean wt yr3
		Pop. Age year 4	Pop. Yr4*(F/Z)*mean wt yr4
		Pop. Age year 5	Pop. Yr5*(F/Z)*mean wt yr5
		Pop. Age year 6	Pop. Yr6*(F/Z)*mean wt yr6
		Total Yield	Sum(Yield.Age yr1:yr6)
47	**Year 2049**		
		Pop. Age year 1	Pop. Yr1*(F/Z)*mean wt yr1
		Pop. Age year 2	Pop. Yr2*(F/Z)*mean wt yr2
		Pop. Age year 3	Pop. Yr3*(F/Z)*mean wt yr3
		Pop. Age year 4	Pop. Yr4*(F/Z)*mean wt yr4
		Pop. Age year 5	Pop. Yr5*(F/Z)*mean wt yr5
		Pop. Age year 6	Pop. Yr6*(F/Z)*mean wt yr6
		Total Yield	Sum(Yield.Age yr1:yr6)
48	**Year 2050**		
		Pop. Age year 1	Pop. Yr1*(F/Z)*mean wt yr1
		Pop. Age year 2	Pop. Yr2*(F/Z)*mean wt yr2
		Pop. Age year 3	Pop. Yr3*(F/Z)*mean wt yr3
		Pop. Age year 4	Pop. Yr4*(F/Z)*mean wt yr4
		Pop. Age year 5	Pop. Yr5*(F/Z)*mean wt yr5
		Pop. Age year 6	Pop. Yr6*(F/Z)*mean wt yr6
		Total Yield	Sum(Yield.Age yr1:yr6)

49	Year 2051		
		Pop. Age year 1	Pop. Yr1*(F/Z)*mean wt yr1
		Pop. Age year 2	Pop. Yr2*(F/Z)*mean wt yr2
		Pop. Age year 3	Pop. Yr3*(F/Z)*mean wt yr3
		Pop. Age year 4	Pop. Yr4*(F/Z)*mean wt yr4
		Pop. Age year 5	Pop. Yr5*(F/Z)*mean wt yr5
		Pop. Age year 6	Pop. Yr6*(F/Z)*mean wt yr6
		Total Yield	Sum(Yield.Age yr1:yr6)
50	Year 2052		
		Pop. Age year 1	Pop. Yr1*(F/Z)*mean wt yr1
		Pop. Age year 2	Pop. Yr2*(F/Z)*mean wt yr2
		Pop. Age year 3	Pop. Yr3*(F/Z)*mean wt yr3
		Pop. Age year 4	Pop. Yr4*(F/Z)*mean wt yr4
		Pop. Age year 5	Pop. Yr5*(F/Z)*mean wt yr5
		Pop. Age year 6	Pop. Yr6*(F/Z)*mean wt yr6
		Total Yield	Sum(Yield.Age yr1:yr6)

6.3.4 Value of Stock Model

The assumed the price per kg was RM 5.00 the resources for this species value in 2003 was RM 791.33 after six modes through natural mortality and fishing mortality the total value RM 3775.45. In 50 years later or 2052 the biomass stock at first mode RM 2.0×10^{-7} and through recruitment, natural mortality (M) and Fishing mortality the total value at six modes RM 2.2×10^{-6} as shown in Table 6.7 and Figure 6. 7 The fallen of biomass stock due to the low recruitment, natural mortality and fishing mortality.

Table 6.7 The value in Ringgit Malaysia (RM) of *Priacanthus tayenus* for 50 Years from 2003 to 2052

Mean Wt	33.39825283	104.6716043	178.9129963	237.9173156	279.2523536	306.358785	
Year	1	2	3	4	5	6	Total
2003	791.3316284	1132.339826	883.6960578	536.5376537	287.5306347	144.0226327	3775.4584348
2004	162.0211112	1132.339826	883.6960578	516.0154691	276.5327918	138.5138692	3109.1191263
2005	159.5570172	231.8407991	883.6960578	516.0154691	265.9556087	133.2158119	2190.2807636
2006	158.8141018	228.3148546	180.9322561	516.0154691	265.9556087	128.1204016	1478.1526918
2007	81.60057636	227.2517949	178.1805521	105.6515337	265.9556087	128.1204016	986.7604670
2008	47.28473426	116.7646779	177.3509244	104.0447346	54.45305326	128.1204016	628.0185252
2009	32.10360806	67.66112453	91.12501634	103.5602904	53.62490536	26.23199826	374.3069426
2010	22.48115964	45.93800205	52.80382056	53.21051017	53.37522175	25.83304957	253.6417634

HAMID AWONG FISHERIES MODEL (HAFM)

Year							
2011	14.0447139	32.1689560	35.8507493	30.8336650	27.4248244	25.7127679	166.0356764
2012	8.4723239	20.0969964	25.1051662	20.9342806	15.8917448	13.2115263	103.7120382
2013	5.6439222	12.1232988	15.6840165	14.6596265	10.7895783	7.6556262	66.5560685
2014	3.7187090	8.0760552	9.4612157	9.1583470	7.5556066	5.1977287	43.1676622
2015	2.3437586	5.3212107	6.3026822	5.5246752	4.7202340	3.6398080	27.8523687
2016	1.4912394	3.3537534	4.1527577	3.6803169	2.8474308	2.2739068	17.7994050
2017	0.9672998	2.1338585	2.6173227	2.4249143	1.8968441	1.3717100	11.4119495
2018	0.6256655	1.3841379	1.6652972	1.5283298	1.2498066	0.9137782	7.3670152
2019	0.3999202	0.8952832	1.0802033	0.9724148	0.7877048	0.6020769	4.7376032
2020	0.2562062	0.5722577	0.6986933	0.6307617	0.5011849	0.3794658	3.0385696
2021	0.1652782	0.3666130	0.4465990	0.4079871	0.3250961	0.2414388	1.9530122
2022	0.1063808	0.2365015	0.2861107	0.2607820	0.2102775	0.1566105	1.2566629
2023	0.0682452	0.1522235	0.1845695	0.1670682	0.1344076	0.1012983	0.8078123
2024	0.0438402	0.0976541	0.1187976	0.1077754	0.0861074	0.0647490	0.5189238
2025	0.0282092	0.0627323	0.0762108	0.0693693	0.0555477	0.0414810	0.3335504
2026	0.0181377	0.0403654	0.0489573	0.0445017	0.0357531	0.0267593	0.2144745
2027	0.0116514	0.0259537	0.0315018	0.0285876	0.0229363	0.0172236	0.1378543
2028	0.0074883	0.0166723	0.0202547	0.0183948	0.0147341	0.0110492	0.0885934
2029	0.0048150	0.0107152	0.0130113	0.0118273	0.0094807	0.0070979	0.0569475
2030	0.0030951	0.0068899	0.0083623	0.0075977	0.0060958	0.0045672	0.0366080
2031	0.0019891	0.0044288	0.0053770	0.0048830	0.0039159	0.0029366	0.0235303
2032	0.0012785	0.0028462	0.0034563	0.0031398	0.0025167	0.0018864	0.0151240
2033	0.0008219	0.0018295	0.0022212	0.0020183	0.0016182	0.0012124	0.0097215
2034	0.0005283	0.0011761	0.0014277	0.0012970	0.0010402	0.0007796	0.0062489
2035	0.0003396	0.0007559	0.0009178	0.0008337	0.0006685	0.0005011	0.0040166
2036	0.0002183	0.0004859	0.0005900	0.0005359	0.0004297	0.0003220	0.0025818
2037	0.0001403	0.0003123	0.0003792	0.0003445	0.0002762	0.0002070	0.0016595
2038	0.0000902	0.0002008	0.0002437	0.0002214	0.0001776	0.0001331	0.0010667
2039	0.0000580	0.0001290	0.0001567	0.0001423	0.0001141	0.0000855	0.0006856
2040	0.0000373	0.0000829	0.0001007	0.0000915	0.0000734	0.0000550	0.0004407
2041	0.0000239	0.0000533	0.0000647	0.0000588	0.0000472	0.0000353	0.0002833
2042	0.0000154	0.0000343	0.0000416	0.0000378	0.0000303	0.0000227	0.0001821
2043	0.0000099	0.0000220	0.0000267	0.0000243	0.0000195	0.0000146	0.0001170
2044	0.0000064	0.0000142	0.0000172	0.0000156	0.0000125	0.0000094	0.0000752
2045	0.0000041	0.0000091	0.0000110	0.0000100	0.0000080	0.0000060	0.0000484
2046	0.0000026	0.0000058	0.0000071	0.0000065	0.0000052	0.0000039	0.0000311
2047	0.0000017	0.0000038	0.0000046	0.0000041	0.0000033	0.0000025	0.0000200
2048	0.0000011	0.0000024	0.0000029	0.0000027	0.0000021	0.0000016	0.0000128
2049	0.0000007	0.0000016	0.0000019	0.0000017	0.0000014	0.0000010	0.0000083
2050	0.0000004	0.0000010	0.0000012	0.0000011	0.0000009	0.0000007	0.0000053
2051	0.0000003	0.0000006	0.0000008	0.0000007	0.0000006	0.0000004	0.0000034
2052	0.0000002	0.0000004	0.0000005	0.0000005	0.0000004	0.0000003	0.0000022

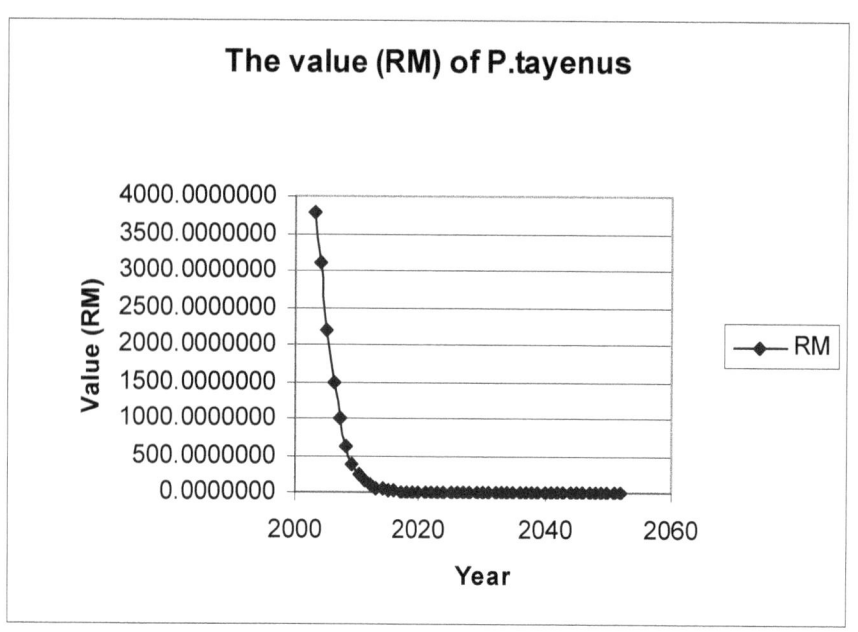

Figure 6.7 The value HAFM Model in Ringgit Malaysia (RM) of *Priacanthus tayenus* for 50 Years from 2003 to 2052

Design HAFM Model The Value Of The Stock

Weight Mean

Age year 1   Winf*(1-Exp(-K*(yr1+to)))^b
Age year 2   Winf*(1-Exp(-K*(yr1+to)))^b
Age year 3   Winf*(1-Exp(-K*(yr1+to)))^b
Age year 4   Winf*(1-Exp(-K*(yr1+to)))^b
Age year 5   Winf*(1-Exp(-K*(yr1+to)))^b
Age year 6   Winf*(1-Exp(-K*(yr1+to)))^b

1Year 2003

Value(RM) Age yr1   Pop. Yr1*(F/Z)*mean wt yr1*price RM5
Value(RM) Age yr2   Pop. Yr2*(F/Z)*mean wt yr2*price RM5
Value(RM) Age yr3   Pop. Yr3*(F/Z)*mean wt yr3*price RM5
Value(RM) Age yr4   Pop. Yr4*(F/Z)*mean wt yr4*price RM5
Value(RM) Age yr5   Pop. Yr5*(F/Z)*mean wt yr5*price RM5
Value(RM) Age yr6   Pop. Yr6*(F/Z)*mean wt yr6*price RM5
Total Value (RM)   Sum(Yield value yr1:yr6)

2Year 2004

Value(RM) Age yr1   Pop. Yr1*(F/Z)*mean wt yr1*price RM5
Value(RM) Age yr2   Pop. Yr2*(F/Z)*mean wt yr2*price RM5
Value(RM) Age yr3   Pop. Yr3*(F/Z)*mean wt yr3*price RM5
Value(RM) Age yr4   Pop. Yr4*(F/Z)*mean wt yr4*price RM5
Value(RM) Age yr5   Pop. Yr5*(F/Z)*mean wt yr5*price RM5
Value(RM) Age yr6   Pop. Yr6*(F/Z)*mean wt yr6*price RM5
Total Value (RM)   Sum(Yield value yr1:yr6)

HAMID AWONG FISHERIES MODEL (HAFM)

3 Year 2005

    Value(RM) Age yr1 Pop. Yr1*(F/Z)*mean wt yr1*price RM5
    Value(RM) Age yr2 Pop. Yr2*(F/Z)*mean wt yr2*price RM5
    Value(RM) Age yr3 Pop. Yr3*(F/Z)*mean wt yr3*price RM5
    Value(RM) Age yr4 Pop. Yr4*(F/Z)*mean wt yr4*price RM5
    Value(RM) Age yr5 Pop. Yr5*(F/Z)*mean wt yr5*price RM5
    Value(RM) Age yr6 Pop. Yr6*(F/Z)*mean wt yr6*price RM5
    Total Value (RM)   Sum(Yield value yr1:yr6)

4 Year 2006

    Value(RM) Age yr1 Pop. Yr1*(F/Z)*mean wt yr1*price RM5
    Value(RM) Age yr2 Pop. Yr2*(F/Z)*mean wt yr2*price RM5
    Value(RM) Age yr3 Pop. Yr3*(F/Z)*mean wt yr3*price RM5
    Value(RM) Age yr4 Pop. Yr4*(F/Z)*mean wt yr4*price RM5
    Value(RM) Age yr5 Pop. Yr5*(F/Z)*mean wt yr5*price RM5
    Value(RM) Age yr6 Pop. Yr6*(F/Z)*mean wt yr6*price RM5
    Total Value (RM)   Sum(Yield value yr1:yr6)

5 Year 2007

    Value(RM) Age yr1 Pop. Yr1*(F/Z)*mean wt yr1*price RM5
    Value(RM) Age yr2 Pop. Yr2*(F/Z)*mean wt yr2*price RM5
    Value(RM) Age yr3 Pop. Yr3*(F/Z)*mean wt yr3*price RM5
    Value(RM) Age yr4 Pop. Yr4*(F/Z)*mean wt yr4*price RM5
    Value(RM) Age yr5 Pop. Yr5*(F/Z)*mean wt yr5*price RM5
    Value(RM) Age yr6 Pop. Yr6*(F/Z)*mean wt yr6*price RM5
    Total Value (RM)   Sum(Yield value yr1:yr6)

6 Year 2008

    Value(RM) Age yr1 Pop. Yr1*(F/Z)*mean wt yr1*price RM5
    Value(RM) Age yr2 Pop. Yr2*(F/Z)*mean wt yr2*price RM5
    Value(RM) Age yr3 Pop. Yr3*(F/Z)*mean wt yr3*price RM5
    Value(RM) Age yr4 Pop. Yr4*(F/Z)*mean wt yr4*price RM5
    Value(RM) Age yr5 Pop. Yr5*(F/Z)*mean wt yr5*price RM5
    Value(RM) Age yr6 Pop. Yr6*(F/Z)*mean wt yr6*price RM5
    Total Value (RM)   Sum(Yield value yr1:yr6)

7 Year 2009

    Value(RM) Age yr1 Pop. Yr1*(F/Z)*mean wt yr1*price RM5
    Value(RM) Age yr2 Pop. Yr2*(F/Z)*mean wt yr2*price RM5
    Value(RM) Age yr3 Pop. Yr3*(F/Z)*mean wt yr3*price RM5
    Value(RM) Age yr4 Pop. Yr4*(F/Z)*mean wt yr4*price RM5
    Value(RM) Age yr5 Pop. Yr5*(F/Z)*mean wt yr5*price RM5
    Value(RM) Age yr6 Pop. Yr6*(F/Z)*mean wt yr6*price RM5
    Total Value (RM)   Sum(Yield value yr1:yr6)

8 Year 2010

    Value(RM) Age yr1 Pop. Yr1*(F/Z)*mean wt yr1*price RM5
    Value(RM) Age yr2 Pop. Yr2*(F/Z)*mean wt yr2*price RM5
    Value(RM) Age yr3 Pop. Yr3*(F/Z)*mean wt yr3*price RM5
    Value(RM) Age yr4 Pop. Yr4*(F/Z)*mean wt yr4*price RM5
    Value(RM) Age yr5 Pop. Yr5*(F/Z)*mean wt yr5*price RM5
    Value(RM) Age yr6 Pop. Yr6*(F/Z)*mean wt yr6*price RM5
    Total Value (RM)   Sum(Yield value yr1:yr6)

9 Year 2011

    Value(RM) Age yr1 Pop. Yr1*(F/Z)*mean wt yr1*price RM5
    Value(RM) Age yr2 Pop. Yr2*(F/Z)*mean wt yr2*price RM5
    Value(RM) Age yr3 Pop. Yr3*(F/Z)*mean wt yr3*price RM5
    Value(RM) Age yr4 Pop. Yr4*(F/Z)*mean wt yr4*price RM5

  Value(RM) Age yr5 Pop. Yr5*(F/Z)*mean wt yr5*price RM5
  Value(RM) Age yr6 Pop. Yr6*(F/Z)*mean wt yr6*price RM5
  Total Value (RM) Sum(Yield value yr1:yr6)

10 Year 2012

  Value(RM) Age yr1 Pop. Yr1*(F/Z)*mean wt yr1*price RM5
  Value(RM) Age yr2 Pop. Yr2*(F/Z)*mean wt yr2*price RM5
  Value(RM) Age yr3 Pop. Yr3*(F/Z)*mean wt yr3*price RM5
  Value(RM) Age yr4 Pop. Yr4*(F/Z)*mean wt yr4*price RM5
  Value(RM) Age yr5 Pop. Yr5*(F/Z)*mean wt yr5*price RM5
  Value(RM) Age yr6 Pop. Yr6*(F/Z)*mean wt yr6*price RM5
  Total Value (RM) Sum(Yield value yr1:yr6)

11 Year 2013

  Value(RM) Age yr1 Pop. Yr1*(F/Z)*mean wt yr1*price RM5
  Value(RM) Age yr2 Pop. Yr2*(F/Z)*mean wt yr2*price RM5
  Value(RM) Age yr3 Pop. Yr3*(F/Z)*mean wt yr3*price RM5
  Value(RM) Age yr4 Pop. Yr4*(F/Z)*mean wt yr4*price RM5
  Value(RM) Age yr5 Pop. Yr5*(F/Z)*mean wt yr5*price RM5
  Value(RM) Age yr6 Pop. Yr6*(F/Z)*mean wt yr6*price RM5
  Total Value (RM) Sum(Yield value yr1:yr6)

12 Year 2014

  Value(RM) Age yr1 Pop. Yr1*(F/Z)*mean wt yr1*price RM5
  Value(RM) Age yr2 Pop. Yr2*(F/Z)*mean wt yr2*price RM5
  Value(RM) Age yr3 Pop. Yr3*(F/Z)*mean wt yr3*price RM5
  Value(RM) Age yr4 Pop. Yr4*(F/Z)*mean wt yr4*price RM5
  Value(RM) Age yr5 Pop. Yr5*(F/Z)*mean wt yr5*price RM5
  Value(RM) Age yr6 Pop. Yr6*(F/Z)*mean wt yr6*price RM5
  Total Value (RM) Sum(Yield value yr1:yr6)

13 Year 2015

  Value(RM) Age yr1 Pop. Yr1*(F/Z)*mean wt yr1*price RM5
  Value(RM) Age yr2 Pop. Yr2*(F/Z)*mean wt yr2*price RM5
  Value(RM) Age yr3 Pop. Yr3*(F/Z)*mean wt yr3*price RM5
  Value(RM) Age yr4 Pop. Yr4*(F/Z)*mean wt yr4*price RM5
  Value(RM) Age yr5 Pop. Yr5*(F/Z)*mean wt yr5*price RM5
  Value(RM) Age yr6 Pop. Yr6*(F/Z)*mean wt yr6*price RM5
  Total Value (RM) Sum(Yield value yr1:yr6)

14 Year 2016

  Value(RM) Age yr1 Pop. Yr1*(F/Z)*mean wt yr1*price RM5
  Value(RM) Age yr2 Pop. Yr2*(F/Z)*mean wt yr2*price RM5
  Value(RM) Age yr3 Pop. Yr3*(F/Z)*mean wt yr3*price RM5
  Value(RM) Age yr4 Pop. Yr4*(F/Z)*mean wt yr4*price RM5
  Value(RM) Age yr5 Pop. Yr5*(F/Z)*mean wt yr5*price RM5
  Value(RM) Age yr6 Pop. Yr6*(F/Z)*mean wt yr6*price RM5
  Total Value (RM) Sum(Yield value yr1:yr6)

15 Year 2017

  Value(RM) Age yr1 Pop. Yr1*(F/Z)*mean wt yr1*price RM5
  Value(RM) Age yr2 Pop. Yr2*(F/Z)*mean wt yr2*price RM5
  Value(RM) Age yr3 Pop. Yr3*(F/Z)*mean wt yr3*price RM5
  Value(RM) Age yr4 Pop. Yr4*(F/Z)*mean wt yr4*price RM5
  Value(RM) Age yr5 Pop. Yr5*(F/Z)*mean wt yr5*price RM5
  Value(RM) Age yr6 Pop. Yr6*(F/Z)*mean wt yr6*price RM5
  Total Value (RM) Sum(Yield value yr1:yr6)

16 Year 2018

HAMID AWONG FISHERIES MODEL (HAFM)

    Value(RM) Age yr1 Pop. Yr1*(F/Z)*mean wt yr1*price RM5
    Value(RM) Age yr2 Pop. Yr2*(F/Z)*mean wt yr2*price RM5
    Value(RM) Age yr3 Pop. Yr3*(F/Z)*mean wt yr3*price RM5
    Value(RM) Age yr4 Pop. Yr4*(F/Z)*mean wt yr4*price RM5
    Value(RM) Age yr5 Pop. Yr5*(F/Z)*mean wt yr5*price RM5
    Value(RM) Age yr6 Pop. Yr6*(F/Z)*mean wt yr6*price RM5
    Total Value (RM)   Sum(Yield value yr1:yr6)

17Year 2019

    Value(RM) Age yr1 Pop. Yr1*(F/Z)*mean wt yr1*price RM5
    Value(RM) Age yr2 Pop. Yr2*(F/Z)*mean wt yr2*price RM5
    Value(RM) Age yr3 Pop. Yr3*(F/Z)*mean wt yr3*price RM5
    Value(RM) Age yr4 Pop. Yr4*(F/Z)*mean wt yr4*price RM5
    Value(RM) Age yr5 Pop. Yr5*(F/Z)*mean wt yr5*price RM5
    Value(RM) Age yr6 Pop. Yr6*(F/Z)*mean wt yr6*price RM5
    Total Value (RM)   Sum(Yield value yr1:yr6)

18Year 2020

    Value(RM) Age yr1 Pop. Yr1*(F/Z)*mean wt yr1*price RM5
    Value(RM) Age yr2 Pop. Yr2*(F/Z)*mean wt yr2*price RM5
    Value(RM) Age yr3 Pop. Yr3*(F/Z)*mean wt yr3*price RM5
    Value(RM) Age yr4 Pop. Yr4*(F/Z)*mean wt yr4*price RM5
    Value(RM) Age yr5 Pop. Yr5*(F/Z)*mean wt yr5*price RM5
    Value(RM) Age yr6 Pop. Yr6*(F/Z)*mean wt yr6*price RM5
    Total Value (RM)   Sum(Yield value yr1:yr6)

19Year 2021

    Value(RM) Age yr1 Pop. Yr1*(F/Z)*mean wt yr1*price RM5
    Value(RM) Age yr2 Pop. Yr2*(F/Z)*mean wt yr2*price RM5
    Value(RM) Age yr3 Pop. Yr3*(F/Z)*mean wt yr3*price RM5
    Value(RM) Age yr4 Pop. Yr4*(F/Z)*mean wt yr4*price RM5
    Value(RM) Age yr5 Pop. Yr5*(F/Z)*mean wt yr5*price RM5
    Value(RM) Age yr6 Pop. Yr6*(F/Z)*mean wt yr6*price RM5
    Total Value (RM)   Sum(Yield value yr1:yr6)

20Year 2022

    Value(RM) Age yr1 Pop. Yr1*(F/Z)*mean wt yr1*price RM5
    Value(RM) Age yr2 Pop. Yr2*(F/Z)*mean wt yr2*price RM5
    Value(RM) Age yr3 Pop. Yr3*(F/Z)*mean wt yr3*price RM5
    Value(RM) Age yr4 Pop. Yr4*(F/Z)*mean wt yr4*price RM5
    Value(RM) Age yr5 Pop. Yr5*(F/Z)*mean wt yr5*price RM5
    Value(RM) Age yr6 Pop. Yr6*(F/Z)*mean wt yr6*price RM5
    Total Value (RM)   Sum(Yield value yr1:yr6)

21Year 2023

    Value(RM) Age yr1 Pop. Yr1*(F/Z)*mean wt yr1*price RM5
    Value(RM) Age yr2 Pop. Yr2*(F/Z)*mean wt yr2*price RM5
    Value(RM) Age yr3 Pop. Yr3*(F/Z)*mean wt yr3*price RM5
    Value(RM) Age yr4 Pop. Yr4*(F/Z)*mean wt yr4*price RM5
    Value(RM) Age yr5 Pop. Yr5*(F/Z)*mean wt yr5*price RM5
    Value(RM) Age yr6 Pop. Yr6*(F/Z)*mean wt yr6*price RM5
    Total Value (RM)   Sum(Yield value yr1:yr6)

22Year 2024

    Value(RM) Age yr1 Pop. Yr1*(F/Z)*mean wt yr1*price RM5
    Value(RM) Age yr2 Pop. Yr2*(F/Z)*mean wt yr2*price RM5
    Value(RM) Age yr3 Pop. Yr3*(F/Z)*mean wt yr3*price RM5
    Value(RM) Age yr4 Pop. Yr4*(F/Z)*mean wt yr4*price RM5
    Value(RM) Age yr5 Pop. Yr5*(F/Z)*mean wt yr5*price RM5

	Value(RM) Age yr6	Pop. Yr6*(F/Z)*mean wt yr6*price RM5
	Total Value (RM)	Sum(Yield value yr1:yr6)

23 Year 2025

    Value(RM) Age yr1 Pop. Yr1*(F/Z)*mean wt yr1*price RM5
    Value(RM) Age yr2 Pop. Yr2*(F/Z)*mean wt yr2*price RM5
    Value(RM) Age yr3 Pop. Yr3*(F/Z)*mean wt yr3*price RM5
    Value(RM) Age yr4 Pop. Yr4*(F/Z)*mean wt yr4*price RM5
    Value(RM) Age yr5 Pop. Yr5*(F/Z)*mean wt yr5*price RM5
    Value(RM) Age yr6 Pop. Yr6*(F/Z)*mean wt yr6*price RM5
    Total Value (RM)   Sum(Yield value yr1:yr6)

24 Year 2026

    Value(RM) Age yr1 Pop. Yr1*(F/Z)*mean wt yr1*price RM5
    Value(RM) Age yr2 Pop. Yr2*(F/Z)*mean wt yr2*price RM5
    Value(RM) Age yr3 Pop. Yr3*(F/Z)*mean wt yr3*price RM5
    Value(RM) Age yr4 Pop. Yr4*(F/Z)*mean wt yr4*price RM5
    Value(RM) Age yr5 Pop. Yr5*(F/Z)*mean wt yr5*price RM5
    Value(RM) Age yr6 Pop. Yr6*(F/Z)*mean wt yr6*price RM5
    Total Value (RM)   Sum(Yield value yr1:yr6)

25 Year 2027

    Value(RM) Age yr1 Pop. Yr1*(F/Z)*mean wt yr1*price RM5
    Value(RM) Age yr2 Pop. Yr2*(F/Z)*mean wt yr2*price RM5
    Value(RM) Age yr3 Pop. Yr3*(F/Z)*mean wt yr3*price RM5
    Value(RM) Age yr4 Pop. Yr4*(F/Z)*mean wt yr4*price RM5
    Value(RM) Age yr5 Pop. Yr5*(F/Z)*mean wt yr5*price RM5
    Value(RM) Age yr6 Pop. Yr6*(F/Z)*mean wt yr6*price RM5
    Total Value (RM)   Sum(Yield value yr1:yr6)

26 Year 2028

    Value(RM) Age yr1 Pop. Yr1*(F/Z)*mean wt yr1*price RM5
    Value(RM) Age yr2 Pop. Yr2*(F/Z)*mean wt yr2*price RM5
    Value(RM) Age yr3 Pop. Yr3*(F/Z)*mean wt yr3*price RM5
    Value(RM) Age yr4 Pop. Yr4*(F/Z)*mean wt yr4*price RM5
    Value(RM) Age yr5 Pop. Yr5*(F/Z)*mean wt yr5*price RM5
    Value(RM) Age yr6 Pop. Yr6*(F/Z)*mean wt yr6*price RM5
    Total Value (RM)   Sum(Yield value yr1:yr6)

27 Year 2029

    Value(RM) Age yr1 Pop. Yr1*(F/Z)*mean wt yr1*price RM5
    Value(RM) Age yr2 Pop. Yr2*(F/Z)*mean wt yr2*price RM5
    Value(RM) Age yr3 Pop. Yr3*(F/Z)*mean wt yr3*price RM5
    Value(RM) Age yr4 Pop. Yr4*(F/Z)*mean wt yr4*price RM5
    Value(RM) Age yr5 Pop. Yr5*(F/Z)*mean wt yr5*price RM5
    Value(RM) Age yr6 Pop. Yr6*(F/Z)*mean wt yr6*price RM5
    Total Value (RM)   Sum(Yield value yr1:yr6)

28 Year 2030

    Value(RM) Age yr1 Pop. Yr1*(F/Z)*mean wt yr1*price RM5
    Value(RM) Age yr2 Pop. Yr2*(F/Z)*mean wt yr2*price RM5
    Value(RM) Age yr3 Pop. Yr3*(F/Z)*mean wt yr3*price RM5
    Value(RM) Age yr4 Pop. Yr4*(F/Z)*mean wt yr4*price RM5
    Value(RM) Age yr5 Pop. Yr5*(F/Z)*mean wt yr5*price RM5
    Value(RM) Age yr6 Pop. Yr6*(F/Z)*mean wt yr6*price RM5
    Total Value (RM)   Sum(Yield value yr1:yr6)

29 Year 2031

    Value(RM) Age yr1 Pop. Yr1*(F/Z)*mean wt yr1*price RM5

HAMID AWONG FISHERIES MODEL (HAFM)

      Value(RM) Age yr2 Pop. Yr2*(F/Z)*mean wt yr2*price RM5
      Value(RM) Age yr3 Pop. Yr3*(F/Z)*mean wt yr3*price RM5
      Value(RM) Age yr4 Pop. Yr4*(F/Z)*mean wt yr4*price RM5
      Value(RM) Age yr5 Pop. Yr5*(F/Z)*mean wt yr5*price RM5
      Value(RM) Age yr6 Pop. Yr6*(F/Z)*mean wt yr6*price RM5
      Total Value (RM) Sum(Yield value yr1:yr6)

30Year 2032

      Value(RM) Age yr1 Pop. Yr1*(F/Z)*mean wt yr1*price RM5
      Value(RM) Age yr2 Pop. Yr2*(F/Z)*mean wt yr2*price RM5
      Value(RM) Age yr3 Pop. Yr3*(F/Z)*mean wt yr3*price RM5
      Value(RM) Age yr4 Pop. Yr4*(F/Z)*mean wt yr4*price RM5
      Value(RM) Age yr5 Pop. Yr5*(F/Z)*mean wt yr5*price RM5
      Value(RM) Age yr6 Pop. Yr6*(F/Z)*mean wt yr6*price RM5
      Total Value (RM) Sum(Yield value yr1:yr6)

31Year 2033

      Value(RM) Age yr1 Pop. Yr1*(F/Z)*mean wt yr1*price RM5
      Value(RM) Age yr2 Pop. Yr2*(F/Z)*mean wt yr2*price RM5
      Value(RM) Age yr3 Pop. Yr3*(F/Z)*mean wt yr3*price RM5
      Value(RM) Age yr4 Pop. Yr4*(F/Z)*mean wt yr4*price RM5
      Value(RM) Age yr5 Pop. Yr5*(F/Z)*mean wt yr5*price RM5
      Value(RM) Age yr6 Pop. Yr6*(F/Z)*mean wt yr6*price RM5
      Total Value (RM) Sum(Yield value yr1:yr6)

32Year 2034

      Value(RM) Age yr1 Pop. Yr1*(F/Z)*mean wt yr1*price RM5
      Value(RM) Age yr2 Pop. Yr2*(F/Z)*mean wt yr2*price RM5
      Value(RM) Age yr3 Pop. Yr3*(F/Z)*mean wt yr3*price RM5
      Value(RM) Age yr4 Pop. Yr4*(F/Z)*mean wt yr4*price RM5
      Value(RM) Age yr5 Pop. Yr5*(F/Z)*mean wt yr5*price RM5
      Value(RM) Age yr6 Pop. Yr6*(F/Z)*mean wt yr6*price RM5
      Total Value (RM) Sum(Yield value yr1:yr6)

33Year 2035

      Value(RM) Age yr1 Pop. Yr1*(F/Z)*mean wt yr1*price RM5
      Value(RM) Age yr2 Pop. Yr2*(F/Z)*mean wt yr2*price RM5
      Value(RM) Age yr3 Pop. Yr3*(F/Z)*mean wt yr3*price RM5
      Value(RM) Age yr4 Pop. Yr4*(F/Z)*mean wt yr4*price RM5
      Value(RM) Age yr5 Pop. Yr5*(F/Z)*mean wt yr5*price RM5
      Value(RM) Age yr6 Pop. Yr6*(F/Z)*mean wt yr6*price RM5
      Total Value (RM) Sum(Yield value yr1:yr6)

34Year 2036

      Value(RM) Age yr1 Pop. Yr1*(F/Z)*mean wt yr1*price RM5
      Value(RM) Age yr2 Pop. Yr2*(F/Z)*mean wt yr2*price RM5
      Value(RM) Age yr3 Pop. Yr3*(F/Z)*mean wt yr3*price RM5
      Value(RM) Age yr4 Pop. Yr4*(F/Z)*mean wt yr4*price RM5
      Value(RM) Age yr5 Pop. Yr5*(F/Z)*mean wt yr5*price RM5
      Value(RM) Age yr6 Pop. Yr6*(F/Z)*mean wt yr6*price RM5
      Total Value (RM) Sum(Yield value yr1:yr6)

35Year 2037

      Value(RM) Age yr1 Pop. Yr1*(F/Z)*mean wt yr1*price RM5
      Value(RM) Age yr2 Pop. Yr2*(F/Z)*mean wt yr2*price RM5
      Value(RM) Age yr3 Pop. Yr3*(F/Z)*mean wt yr3*price RM5
      Value(RM) Age yr4 Pop. Yr4*(F/Z)*mean wt yr4*price RM5
      Value(RM) Age yr5 Pop. Yr5*(F/Z)*mean wt yr5*price RM5
      Value(RM) Age yr6 Pop. Yr6*(F/Z)*mean wt yr6*price RM5

Total Value (RM)   Sum(Yield value yr1:yr6)

36 Year 2038

Value(RM) Age yr1 Pop. Yr1*(F/Z)*mean wt yr1*price RM5
Value(RM) Age yr2 Pop. Yr2*(F/Z)*mean wt yr2*price RM5
Value(RM) Age yr3 Pop. Yr3*(F/Z)*mean wt yr3*price RM5
Value(RM) Age yr4 Pop. Yr4*(F/Z)*mean wt yr4*price RM5
Value(RM) Age yr5 Pop. Yr5*(F/Z)*mean wt yr5*price RM5
Value(RM) Age yr6 Pop. Yr6*(F/Z)*mean wt yr6*price RM5
Total Value (RM)   Sum(Yield value yr1:yr6)

37 Year 2039

Value(RM) Age yr1 Pop. Yr1*(F/Z)*mean wt yr1*price RM5
Value(RM) Age yr2 Pop. Yr2*(F/Z)*mean wt yr2*price RM5
Value(RM) Age yr3 Pop. Yr3*(F/Z)*mean wt yr3*price RM5
Value(RM) Age yr4 Pop. Yr4*(F/Z)*mean wt yr4*price RM5
Value(RM) Age yr5 Pop. Yr5*(F/Z)*mean wt yr5*price RM5
Value(RM) Age yr6 Pop. Yr6*(F/Z)*mean wt yr6*price RM5
Total Value (RM)   Sum(Yield value yr1:yr6)

38 Year 2040

Value(RM) Age yr1 Pop. Yr1*(F/Z)*mean wt yr1*price RM5
Value(RM) Age yr2 Pop. Yr2*(F/Z)*mean wt yr2*price RM5
Value(RM) Age yr3 Pop. Yr3*(F/Z)*mean wt yr3*price RM5
Value(RM) Age yr4 Pop. Yr4*(F/Z)*mean wt yr4*price RM5
Value(RM) Age yr5 Pop. Yr5*(F/Z)*mean wt yr5*price RM5
Value(RM) Age yr6 Pop. Yr6*(F/Z)*mean wt yr6*price RM5
Total Value (RM)   Sum(Yield value yr1:yr6)

39 Year 2041

Value(RM) Age yr1 Pop. Yr1*(F/Z)*mean wt yr1*price RM5
Value(RM) Age yr2 Pop. Yr2*(F/Z)*mean wt yr2*price RM5
Value(RM) Age yr3 Pop. Yr3*(F/Z)*mean wt yr3*price RM5
Value(RM) Age yr4 Pop. Yr4*(F/Z)*mean wt yr4*price RM5
Value(RM) Age yr5 Pop. Yr5*(F/Z)*mean wt yr5*price RM5
Value(RM) Age yr6 Pop. Yr6*(F/Z)*mean wt yr6*price RM5
Total Value (RM)   Sum(Yield value yr1:yr6)

40 Year 2042

Value(RM) Age yr1 Pop. Yr1*(F/Z)*mean wt yr1*price RM5
Value(RM) Age yr2 Pop. Yr2*(F/Z)*mean wt yr2*price RM5
Value(RM) Age yr3 Pop. Yr3*(F/Z)*mean wt yr3*price RM5
Value(RM) Age yr4 Pop. Yr4*(F/Z)*mean wt yr4*price RM5
Value(RM) Age yr5 Pop. Yr5*(F/Z)*mean wt yr5*price RM5
Value(RM) Age yr6 Pop. Yr6*(F/Z)*mean wt yr6*price RM5
Total Value (RM)   Sum(Yield value yr1:yr6)

41 Year 2043

Value(RM) Age yr1 Pop. Yr1*(F/Z)*mean wt yr1*price RM5
Value(RM) Age yr2 Pop. Yr2*(F/Z)*mean wt yr2*price RM5
Value(RM) Age yr3 Pop. Yr3*(F/Z)*mean wt yr3*price RM5
Value(RM) Age yr4 Pop. Yr4*(F/Z)*mean wt yr4*price RM5
Value(RM) Age yr5 Pop. Yr5*(F/Z)*mean wt yr5*price RM5
Value(RM) Age yr6 Pop. Yr6*(F/Z)*mean wt yr6*price RM5
Total Value (RM)   Sum(Yield value yr1:yr6)

42 Year 2044

Value(RM) Age yr1 Pop. Yr1*(F/Z)*mean wt yr1*price RM5
Value(RM) Age yr2 Pop. Yr2*(F/Z)*mean wt yr2*price RM5

HAMID AWONG FISHERIES MODEL (HAFM)

Value(RM) Age yr3 Pop. Yr3*(F/Z)*mean wt yr3*price RM5
Value(RM) Age yr4 Pop. Yr4*(F/Z)*mean wt yr4*price RM5
Value(RM) Age yr5 Pop. Yr5*(F/Z)*mean wt yr5*price RM5
Value(RM) Age yr6 Pop. Yr6*(F/Z)*mean wt yr6*price RM5
Total Value (RM)   Sum(Yield value yr1:yr6)

43Year 2045

Value(RM) Age yr1 Pop. Yr1*(F/Z)*mean wt yr1*price RM5
Value(RM) Age yr2 Pop. Yr2*(F/Z)*mean wt yr2*price RM5
Value(RM) Age yr3 Pop. Yr3*(F/Z)*mean wt yr3*price RM5
Value(RM) Age yr4 Pop. Yr4*(F/Z)*mean wt yr4*price RM5
Value(RM) Age yr5 Pop. Yr5*(F/Z)*mean wt yr5*price RM5
Value(RM) Age yr6 Pop. Yr6*(F/Z)*mean wt yr6*price RM5
Total Value (RM)   Sum(Yield value yr1:yr6)

44Year 2046

Value(RM) Age yr1 Pop. Yr1*(F/Z)*mean wt yr1*price RM5
Value(RM) Age yr2 Pop. Yr2*(F/Z)*mean wt yr2*price RM5
Value(RM) Age yr3 Pop. Yr3*(F/Z)*mean wt yr3*price RM5
Value(RM) Age yr4 Pop. Yr4*(F/Z)*mean wt yr4*price RM5
Value(RM) Age yr5 Pop. Yr5*(F/Z)*mean wt yr5*price RM5
Value(RM) Age yr6 Pop. Yr6*(F/Z)*mean wt yr6*price RM5
Total Value (RM)   Sum(Yield value yr1:yr6)

45Year 2047

Value(RM) Age yr1 Pop. Yr1*(F/Z)*mean wt yr1*price RM5
Value(RM) Age yr2 Pop. Yr2*(F/Z)*mean wt yr2*price RM5
Value(RM) Age yr3 Pop. Yr3*(F/Z)*mean wt yr3*price RM5
Value(RM) Age yr4 Pop. Yr4*(F/Z)*mean wt yr4*price RM5
Value(RM) Age yr5 Pop. Yr5*(F/Z)*mean wt yr5*price RM5
Value(RM) Age yr6 Pop. Yr6*(F/Z)*mean wt yr6*price RM5
Total Value (RM)   Sum(Yield value yr1:yr6)

46Year 2048

Value(RM) Age yr1 Pop. Yr1*(F/Z)*mean wt yr1*price RM5
Value(RM) Age yr2 Pop. Yr2*(F/Z)*mean wt yr2*price RM5
Value(RM) Age yr3 Pop. Yr3*(F/Z)*mean wt yr3*price RM5
Value(RM) Age yr4 Pop. Yr4*(F/Z)*mean wt yr4*price RM5
Value(RM) Age yr5 Pop. Yr5*(F/Z)*mean wt yr5*price RM5
Value(RM) Age yr6 Pop. Yr6*(F/Z)*mean wt yr6*price RM5
Total Value (RM)   Sum(Yield value yr1:yr6)

47Year 2049

Value(RM) Age yr1 Pop. Yr1*(F/Z)*mean wt yr1*price RM5
Value(RM) Age yr2 Pop. Yr2*(F/Z)*mean wt yr2*price RM5
Value(RM) Age yr3 Pop. Yr3*(F/Z)*mean wt yr3*price RM5
Value(RM) Age yr4 Pop. Yr4*(F/Z)*mean wt yr4*price RM5
Value(RM) Age yr5 Pop. Yr5*(F/Z)*mean wt yr5*price RM5
Value(RM) Age yr6 Pop. Yr6*(F/Z)*mean wt yr6*price RM5
Total Value (RM)   Sum(Yield value yr1:yr6)

48Year 2050

Value(RM) Age yr1 Pop. Yr1*(F/Z)*mean wt yr1*price RM5
Value(RM) Age yr2 Pop. Yr2*(F/Z)*mean wt yr2*price RM5
Value(RM) Age yr3 Pop. Yr3*(F/Z)*mean wt yr3*price RM5
Value(RM) Age yr4 Pop. Yr4*(F/Z)*mean wt yr4*price RM5
Value(RM) Age yr5 Pop. Yr5*(F/Z)*mean wt yr5*price RM5
Value(RM) Age yr6 Pop. Yr6*(F/Z)*mean wt yr6*price RM5
Total Value (RM)   Sum(Yield value yr1:yr6)

49 Year 2051

Value(RM) Age yr1	Pop. Yr1*(F/Z)*mean wt yr1*price RM5
Value(RM) Age yr2	Pop. Yr2*(F/Z)*mean wt yr2*price RM5
Value(RM) Age yr3	Pop. Yr3*(F/Z)*mean wt yr3*price RM5
Value(RM) Age yr4	Pop. Yr4*(F/Z)*mean wt yr4*price RM5
Value(RM) Age yr5	Pop. Yr5*(F/Z)*mean wt yr5*price RM5
Value(RM) Age yr6	Pop. Yr6*(F/Z)*mean wt yr6*price RM5
Total Value (RM)	Sum(Yield value yr1:yr6)

50 Year 2052

Value(RM) Age yr1	Pop. Yr1*(F/Z)*mean wt yr1*price RM5
Value(RM) Age yr2	Pop. Yr2*(F/Z)*mean wt yr2*price RM5
Value(RM) Age yr3	Pop. Yr3*(F/Z)*mean wt yr3*price RM5
Value(RM) Age yr4	Pop. Yr4*(F/Z)*mean wt yr4*price RM5
Value(RM) Age yr5	Pop. Yr5*(F/Z)*mean wt yr5*price RM5
Value(RM) Age yr6	Pop. Yr6*(F/Z)*mean wt yr6*price RM5
Total Value (RM)	Sum(Yield value yr1:yr6)

6.3.5 Description on the Model

The sensitivity of the Threadfin big eye (*Priachantus tayenus*) fishery was change the total mortality to 0.8 per year, by reducing total mortality The results indicate that;

a. The Number Parental Stock (Sm)

The assumed the total mortality (Z) was 0.8 per year initials number of parental stock in 2003 was 100 after six modes through natural mortality and fishing mortality the total parental stock (Sm) 15.8475. In 50 years later or 2052 the parental stock at first mode 3.9×10^{-8} and through recruitment, natural mortality (M) and Fishing mortality (F) the total parental stock 8.0×10^{-6} as shown in Table 6.8 and Figure 6.8 The fallen of number of parental stock due to the low recruitment, natural mortality and fishing mortality.

Table 6.8 The HAFM Model number of *Priacanthus tayenus* for 2003 to 2052 (50 Years) with initials number 100 heads and total mortality (Z) at 0.8 per year

Mean Wt	33.39825283	104.6716043	178.9129963	237.9173156	279.2523536	306.358785	
Year	1	2	3	4	5	6	Sm
2003	100.00000000000	45.657604962332	20.846168908963	9.5178614502348	4.3456275818102	1.9841094744370	15.8475985064820
2004	20.9486312926	45.657604962332	20.846168908963	9.3667874816771	4.2766508260452	1.9526163397740	15.5960546474963
2005	20.8164395640	9.564643320596	20.846168908963	9.3667874816771	4.2087689162481	1.9216230855580	15.4971794834831
2006	20.7756590752	9.504287743372	4.3669870633736	9.3667874816771	4.2087689162481	1.8911217773465	15.4666781752717
2007	10.8535275255	9.485668348891	4.3394301523522	1.9622137734990	4.2087689162481	1.8911217773465	8.0621044670936
2008	6.36417003644	4.955460722093	4.3309289827738	1.9498316552155	0.8816794822228	1.8911217773465	4.7226329147848
2009	4.33889361313	2.905727614368	2.2625446805567	1.9460118334950	0.8761158378409	0.3961641284306	3.2182917997665
2010	3.08107739532	1.981034905618	1.3266856354497	1.0166268575835	0.8743994813042	0.3936642218638	2.2846905607514
2011	1.95027609577	1.406746145738	0.9044930913730	0.5961182822858	0.4567998928117	0.3928930131590	1.4458111882566
2012	1.18660123405	0.890449355481	0.6422865980437	0.4064149437978	0.2678532102708	0.2052534226460	0.8795215767146
2013	0.79817689901	0.541773703921	0.4065578491153	0.2885979717653	0.1826140056984	0.1203542055065	0.5915661829702
2014	0.53218982202	0.364428455450	0.2473608975260	0.1826782171967	0.1296754276997	0.0820537620138	0.3944074069101
2015	0.33936818613	0.242985126586	0.1663893045599	0.1111464158485	0.0820826140998	0.0582669255997	0.2514959555480
2016	0.21805099654	0.154947385790	0.1109411892141	0.0747635338581	0.0499413038985	0.0368820959655	0.1615869337221
2017	0.14288400997	0.099556862619	0.0707452653036	0.0498490896275	0.0335934212222	0.0224400743474	0.1058825851971
2018	0.09349281183	0.065237416828	0.0454552790473	0.0317878967750	0.0223986398045	0.0150944971589	0.0692810337385
2019	0.06041894012	0.042686578692	0.0297858420628	0.0204243734480	0.0142832227294	0.0100643576210	0.0447719537984
2020	0.03910650986	0.027585841004	0.0194896694710	0.0133836415595	0.0091772625641	0.0064178656733	0.0289787697968
2021	0.02549803811	0.017855095787	0.0125950343111	0.0087572729944	0.0060136577980	0.0041236098814	0.0188945406738
2022	0.01659382364	0.011641793512	0.0081522091002	0.0056593137200	0.0039348964031	0.0027021106289	0.0122963207520
2023	0.01076086049	0.007576342447	0.0053153640922	0.0036630236703	0.0025428935714	0.0017680629247	0.0079739801664
2024	0.00698614203	0.004913151174	0.0034591765048	0.0023883470415	0.0016459026313	0.0011425957343	0.0051768454071
2025	0.00454378002	0.003189705129	0.0022432271543	0.0015543081956	0.0010731535021	0.0007395517244	0.0033670134221
2026	0.00295343492	0.002074581133	0.0014563429672	0.0010079469335	0.0006983956915	0.0004821989514	0.0021885415764
2027	0.00191775506	0.001348467648	0.0009472040585	0.0006543770768	0.0004528997515	0.0003138094126	0.0014210862409
2028	0.00124577407	0.000875601032	0.0006156780317	0.0004256062184	0.0002940305741	0.0002035009762	0.0009231377687

Year							
2029	0.00080969395	0.000568790605	0.0003997784600	0.0002766419722	0.0001912372012	0.0001321164533	0.0005999956267
2030	0.00052612101	0.000369686866	0.0002596961677	0.0001796320413	0.0001243032508	0.0000859284135	0.0003898637057
2031	0.00034176887	0.000240214252	0.0001687901690	0.0001166890100	0.0000807138790	0.0000558530509	0.0002532559400
2032	0.00022204827	0.000156043478	0.0001096760743	0.0000758423118	0.0000524317520	0.0000362670837	0.0001645411474
2033	0.00014428569	0.000101381921	0.0000712457149	0.0000492806368	0.0000340781474	0.0000235591048	0.0001069178890
2034	0.00009374744	0.000065877389	0.0000462885569	0.0000320127633	0.0000221432175	0.0000153122987	0.0000694682794
2035	0.00006090650	0.000042802836	0.0000300780379	0.0000207987893	0.0000143842618	0.0000099495890	0.0000451326401
2036	0.00003957233	0.000027808448	0.0000195427499	0.0000135149336	0.0000093454985	0.0000064632654	0.0000293236975
2037	0.00002571199	0.000018067778	0.0000126966712	0.0000087811236	0.0000060726511	0.0000041992031	0.0000190529779
2038	0.00001670575	0.000011739480	0.0000082493146	0.0000057049821	0.0000039456132	0.0000027286180	0.0000123792133
2039	0.00001085396	0.000007627445	0.0000053599656	0.0000037066560	0.0000025634137	0.0000017728783	0.0000080429480
2040	0.00000705210	0.000004955657	0.0000034825085	0.0000024083878	0.0000016655079	0.0000011518160	0.0000052257117
2041	0.00000458197	0.000003219819	0.0000022626345	0.0000015647919	0.0000010821584	0.0000007483609	0.0000033953113
2042	0.00000297702	0.000002092019	0.0000014700922	0.0000010166672	0.0000007031063	0.0000004862451	0.0000022060187
2043	0.00000193424	0.000001359237	0.0000009551657	0.0000006605550	0.0000004568180	0.0000003159260	0.0000014332991
2044	0.00000125672	0.000000883126	0.0000006205951	0.0000004291836	0.0000002968065	0.0000002052616	0.0000009312517
2045	0.00000081653	0.000000573790	0.0000004032143	0.0000002788513	0.0000001928446	0.0000001333638	0.0000006050597
2046	0.00000053052	0.000000372807	0.0000002619788	0.0000001811759	0.0000001252960	0.0000000866507	0.0000003931225
2047	0.00000034469	0.000000242222	0.0000001702148	0.0000001177147	0.0000000814076	0.0000000562991	0.0000002554214
2048	0.00000022395	0.000000157378	0.0000001105929	0.0000000764825	0.0000000528926	0.0000000365788	0.0000001659538
2049	0.00000014551	0.000000102252	0.0000000718549	0.0000000496926	0.0000000343658	0.0000000237662	0.0000001078245
2050	0.00000009454	0.000000066436	0.0000000466860	0.0000000322865	0.0000000223283	0.0000000154415	0.0000000700563
2051	0.00000006143	0.000000061143	0.0000000614343	0.0000000614343	0.0000000614343	0.0000000614343	0.0000000455174
2052	0.00000003991	0.000000028046	0.0000000197082	0.0000000136295	0.0000000094257	0.0000000065185	0.0000000295738

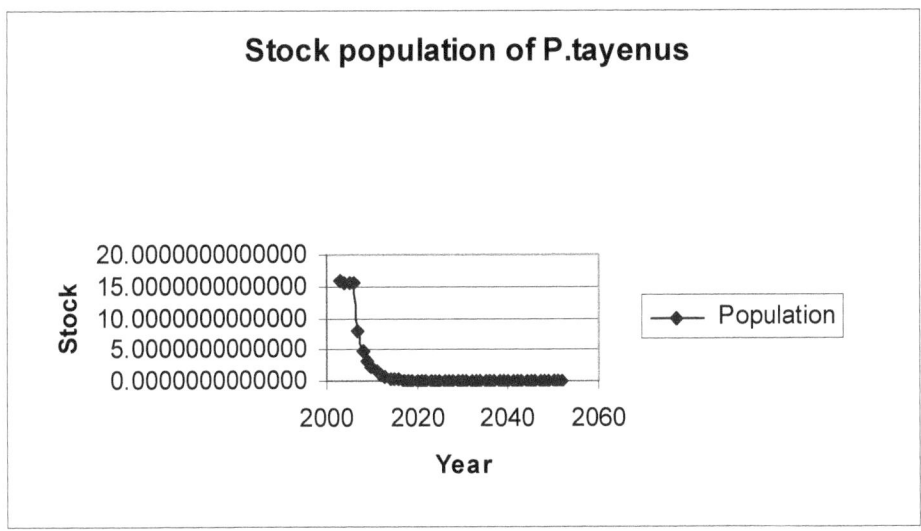

Figura 6.8 The numbers of parental stock used HAFM Model program for recruitment of stock with total mortality at 0.8 per Year.

b. *Biomass (Yield) HAFM Model with Total Mortality (Z) at 0.8 per year*

The assumption assumed of total mortality (Z) was 0.8 per year, the biomass (yield) stock in 2003 was 66.7965057 gram after six modes Through natural mortality and fishing mortality the total biomass 318.6874145 gram. In 50 years later or 2052 the biomass at first mode 1.0×10^{-7} gram and through recruitment, natural mortality (M) and Fishing mortality (F) the yield 3.0×10^{-7} gram as shown in Table 6.9 and Figure 6.9 Reduction number of parental stock due to the low recruitment, natural mortality and fishing mortality.

Table 6.9 The HAFM Model yield (biomass) of *Priacanthus tayenus* for 2003 to 2052 (50 years) with total mortality (z) at 0.8 per year

Mean	33.3982528	104.671604	178.912996	237.917315	279.252353		
Wt	3	3	3	6	6	306.358785	
Year	1	2	3	4	5	6	Total
2003	66.7965057	95.5810952	74.5930108	45.2892809	24.2705346	12.1569874	318.6874145
2004	13.9929537	95.5810952	74.5930108	44.5704187	23.8852962	11.9640234	264.5867979
2005	13.9046542	20.0229312	74.5930108	44.5704187	23.5061725	11.7741223	188.3713097
2006	13.8774143	19.8965809	15.6262148	44.5704187	23.5061725	11.5872354	129.0640366
2007	7.2497771	19.8576025	15.5276090	9.3368927	23.5061725	11.5872354	87.0652892
2008	4.2510432	10.3739205	15.4971896	9.2779743	4.9242214	11.5872354	55.9115844
2009	2.8982293	6.0829434	8.0959730	9.2597982	4.8931482	2.4273672	33.6574593
2010	2.0580520	4.1471620	4.7472260	4.8374627	4.8835623	2.4120499	23.0855149
2011	1.3027163	2.9449275	3.2365114	2.8365372	2.5512489	2.4073245	15.2792658
2012	0.7926082	1.8640953	2.2982684	1.9338630	1.4959728	1.2576238	9.6424314

Year							
2013	0.5331543	1.1341665	1.4547697	1.3732491	1.0199078	0.7374314	6.2526787
2014	0.3554842	0.7629062	0.8851216	0.8692462	0.7242434	0.5027578	4.0997594
2015	0.2266861	0.5086729	0.5953842	0.5288731	0.4584353	0.3570117	2.6750632
2016	0.1456504	0.3243718	0.3969764	0.3557508	0.2789245	0.2259831	1.7276571
2017	0.0954415	0.2084155	0.2531449	0.2371992	0.1876208	0.1374943	1.1193164
2018	0.0624499	0.1365701	0.1626508	0.1512578	0.1250975	0.0924866	0.7305128
2019	0.0403577	0.0893615	0.1065815	0.0971862	0.0797725	0.0616661	0.4749255
2020	0.0261218	0.0577491	0.0697391	0.0636840	0.0512554	0.0393234	0.3078728
2021	0.0170318	0.0373784	0.0450683	0.0416701	0.0335866	0.0252661	0.2000013
2022	0.0110841	0.0243713	0.0291707	0.0269290	0.0219766	0.0165563	0.1300880
2023	0.0071879	0.0158606	0.0190198	0.0174299	0.0142022	0.0108332	0.0845335
2024	0.0046665	0.0102853	0.0123778	0.0113646	0.0091924	0.0070009	0.0548876
2025	0.0030351	0.0066774	0.0080268	0.0073959	0.0059936	0.0045314	0.0356603
2026	0.0019728	0.0043430	0.0052112	0.0047962	0.0039006	0.0029545	0.0231782
2027	0.0012810	0.0028229	0.0033893	0.0031138	0.0025295	0.0019228	0.0150592
2028	0.0008321	0.0018330	0.0022031	0.0020252	0.0016422	0.0012469	0.0097824
2029	0.0005408	0.0011907	0.0014305	0.0013164	0.0010681	0.0008095	0.0063560
2030	0.0003514	0.0007739	0.0009293	0.0008548	0.0006942	0.0005265	0.0041301
2031	0.0002283	0.0005029	0.0006040	0.0005552	0.0004508	0.0003422	0.0026834
2032	0.0001483	0.0003267	0.0003924	0.0003609	0.0002928	0.0002222	0.0017434
2033	0.0000964	0.0002122	0.0002549	0.0002345	0.0001903	0.0001444	0.0011327
2034	0.0000626	0.0001379	0.0001656	0.0001523	0.0001237	0.0000938	0.0007360
2035	0.0000407	0.0000896	0.0001076	0.0000990	0.0000803	0.0000610	0.0004782
2036	0.0000264	0.0000582	0.0000699	0.0000643	0.0000522	0.0000396	0.0003107
2037	0.0000172	0.0000378	0.0000454	0.0000418	0.0000339	0.0000257	0.0002019
2038	0.0000112	0.0000246	0.0000295	0.0000271	0.0000220	0.0000167	0.0001312
2039	0.0000073	0.0000160	0.0000192	0.0000176	0.0000143	0.0000109	0.0000852
2040	0.0000047	0.0000104	0.0000125	0.0000115	0.0000093	0.0000071	0.0000554
2041	0.0000031	0.0000067	0.0000081	0.0000074	0.0000060	0.0000046	0.0000360
2042	0.0000020	0.0000044	0.0000053	0.0000048	0.0000039	0.0000030	0.0000234
2043	0.0000013	0.0000028	0.0000034	0.0000031	0.0000026	0.0000019	0.0000152
2044	0.0000008	0.0000018	0.0000022	0.0000020	0.0000017	0.0000013	0.0000099
2045	0.0000005	0.0000012	0.0000014	0.0000013	0.0000011	0.0000008	0.0000064
2046	0.0000004	0.0000008	0.0000009	0.0000009	0.0000007	0.0000005	0.0000042
2047	0.0000002	0.0000005	0.0000006	0.0000006	0.0000005	0.0000003	0.0000027
2048	0.0000001	0.0000003	0.0000004	0.0000004	0.0000003	0.0000002	0.0000018
2049	0.0000001	0.0000002	0.0000003	0.0000002	0.0000002	0.0000001	0.0000011
2050	0.0000001	0.0000001	0.0000002	0.0000002	0.0000001	0.0000001	0.0000007
2051	0.0000001	0.0000001	0.0000002	0.0000002	0.0000001	0.0000001	0.0000209
2052	0.0000001	0.0000001	0.0000002	0.0000002	0.0000001	0.0000001	0.0000003

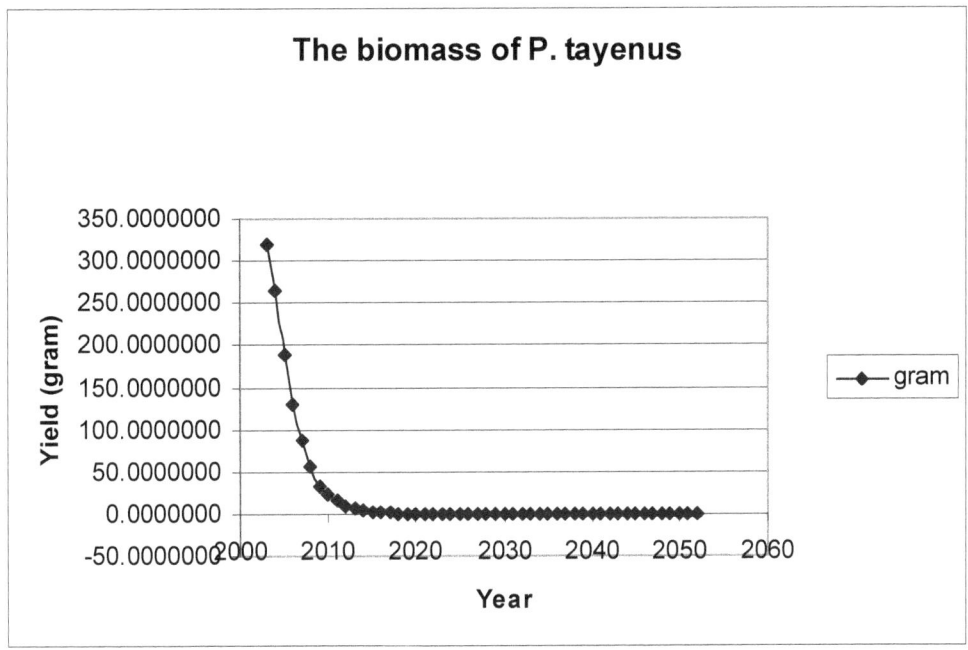

Figura 6.9 The yield (biomass) of stock used HAFM Model program with total mortality at 0.8 per Year

c. *Value Model with total mortality (Z) at 0.8 per year*

Assumed the total mortality (Z) 0.8 per year and the price per kg was RM 5.00 the resources value for this species value in 2003 was RM 333.98. After six modes through natural mortality and fishing mortality the total value RM 1593.43 In 50 years later or 2052 the biomass stock at first mode RM 3.0×10^{-7} and through recruitment, natural mortality (M) and Fishing mortality (F) the total value at six modes RM 1.6×10^{-6} as shown in Table 6.10 and Figure 6.10 The fallen of biomass stock due to the low recruitment, natural mortality and fishing mortality.

Table 6.10 The value in Ringgit Malaysia (RM) of *Priacanthus tayenus* for 2003 to 2052 (50 years) with total mortality (z) at 0.8 per year

Mean	33.3982528	104.671604	178.912996	237.917315	279.252353		
Wt	3	3	3	6	6306.358785		
Year	1	2	3	4	5	6	Total
2003	333.982528	477.905475	372.965054	226.446404	121.352673	60.7849368	1593.437072
2004	69.9647684	477.905475	372.965054	222.852093	119.426480	59.8201169	1322.933989
2005	69.5232712	100.114656	372.965054	222.852093	117.530862	58.8706114	941.8565486

	1	1	4	6			
2006	69.3870715	99.4829046	78.1310740	222.8520934	117.5308626	57.9361770	645.3201830
2007	36.2488856	99.2880124	77.6380451	46.6844634	117.5308626	57.9361770	435.3264460
2008	21.2552160	51.8696024	77.4859481	46.3897133	24.6210710	57.9361770	279.5579219
2009	14.4911466	30.4147171	40.4798648	46.2989912	24.4657410	12.1368361	168.2872967
2010	10.2902602	20.7358102	23.7361302	24.1871330	24.4178113	12.0602493	115.4275745
2011	6.5135814	14.7246376	16.1825569	14.1826862	12.7562445	12.0366226	76.3963292
2012	3.9630408	9.3204763	11.4913420	9.6693152	7.4798639	6.2881189	48.2121571
2013	2.6657714	5.6708323	7.2738483	6.8662455	5.0995391	3.6871568	31.2633933
2014	1.7774210	3.8145311	4.4256079	4.3462311	3.6212168	2.5137891	20.4987971
2015	1.1334304	2.5433643	2.9769209	2.6443657	2.2921763	1.7850585	13.3753161
2016	0.7282522	1.6218591	1.9848821	1.7787539	1.3946227	1.1299154	8.6382854
2017	0.4772076	1.0420777	1.2657247	1.1859962	0.9381042	0.6874714	5.5965818
2018	0.3122497	0.6828505	0.8132540	0.7562891	0.6254873	0.4624332	3.6525638
2019	0.2017887	0.4468073	0.5329074	0.4859312	0.3988624	0.3083304	2.3746274
2020	0.1306089	0.2887454	0.3486955	0.3184200	0.2562772	0.1966170	1.5393640
2021	0.0851590	0.1868922	0.2253415	0.2083507	0.1679328	0.1263304	1.0000066
2022	0.0554205	0.1218565	0.1458536	0.1346449	0.1098829	0.0827815	0.6504399
2023	0.0359394	0.0793028	0.0950988	0.0871497	0.0710109	0.0541662	0.4226677
2024	0.0233325	0.0514267	0.0618892	0.0568229	0.0459622	0.0350044	0.2744380
2025	0.0151754	0.0333872	0.0401342	0.0369797	0.0299681	0.0226568	0.1783014
2026	0.0098640	0.0217150	0.0260559	0.0239808	0.0195029	0.0147726	0.1158911
2027	0.0064050	0.0141146	0.0169467	0.0155688	0.0126473	0.0096138	0.0752962
2028	0.0041607	0.0091651	0.0110153	0.0101259	0.0082109	0.0062344	0.0489122
2029	0.0027042	0.0059536	0.0071526	0.0065818	0.0053403	0.0040475	0.0317801
2030	0.0017572	0.0038696	0.0046463	0.0042738	0.0034712	0.0026325	0.0206505
2031	0.0011414	0.0025144	0.0030199	0.0027762	0.0022540	0.0017111	0.0134170
2032	0.0007416	0.0016333	0.0019622	0.0018044	0.0014642	0.0011111	0.0087168
2033	0.0004819	0.0010612	0.0012747	0.0011725	0.0009516	0.0007218	0.0056636
2034	0.0003131	0.0006895	0.0008282	0.0007616	0.0006184	0.0004691	0.0036799
2035	0.0002034	0.0004480	0.0005381	0.0004948	0.0004017	0.0003048	0.0023909
2036	0.0001322	0.0002911	0.0003496	0.0003215	0.0002610	0.0001980	0.0015534
2037	0.0000859	0.0001891	0.0002272	0.0002089	0.0001696	0.0001286	0.0010093
2038	0.0000558	0.0001229	0.0001476	0.0001357	0.0001102	0.0000836	0.0006558
2039	0.0000363	0.0000798	0.0000959	0.0000882	0.0000716	0.0000543	0.0004261
2040	0.0000236	0.0000519	0.0000623	0.0000573	0.0000465	0.0000353	0.0002768
2041	0.0000153	0.0000337	0.0000405	0.0000372	0.0000302	0.0000229	0.0001799
2042	0.0000099	0.0000219	0.0000263	0.0000242	0.0000196	0.0000149	0.0001169
2043	0.0000065	0.0000142	0.0000171	0.0000157	0.0000128	0.0000097	0.0000759
2044	0.0000042	0.0000092	0.0000111	0.0000102	0.0000083	0.0000063	0.0000493
2045	0.0000027	0.0000060	0.0000072	0.0000066	0.0000054	0.0000041	0.0000321
2046	0.0000018	0.0000039	0.0000047	0.0000043	0.0000035	0.0000027	0.0000208
2047	0.0000012	0.0000025	0.0000030	0.0000028	0.0000023	0.0000017	0.0000135
2048	0.0000007	0.0000016	0.0000020	0.0000018	0.0000015	0.0000011	0.0000088
2049	0.0000005	0.0000011	0.0000013	0.0000012	0.0000010	0.0000007	0.0000057
2050	0.0000003	0.0000007	0.0000008	0.0000008	0.0000006	0.0000005	0.0000037
2051	0.0000003	0.0000007	0.0000008	0.0000008	0.0000006	0.0000005	0.0001046
2052	0.0000003	0.0000007	0.0000008	0.0000008	0.0000006	0.0000005	0.0000016

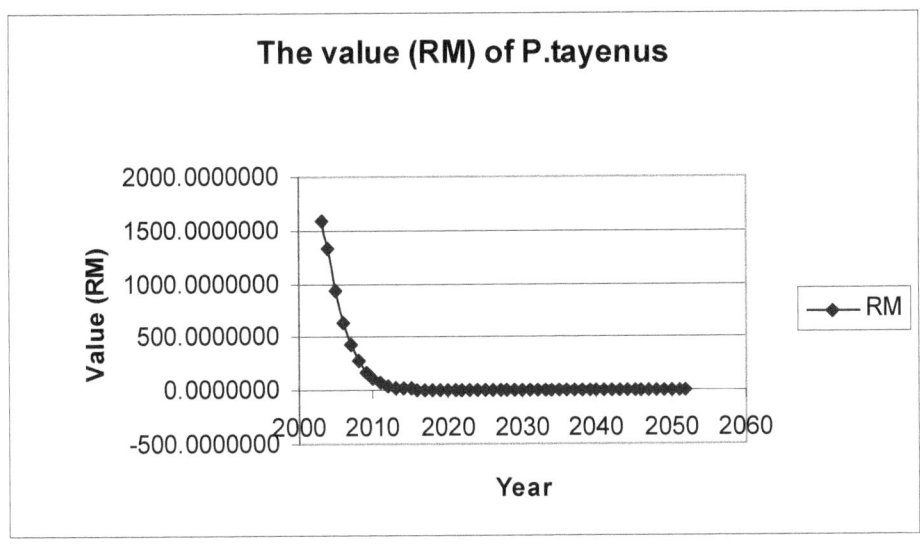

Figura 6.10 The value in Ringgit Malaysia (RM) used Value HAFM Model with total mortality (Z) at 0.8 per year

6.3.6 Management Regime

Based on the above discussion, input and output restriction could be imposed as TAC in Threadfin big eye (*Priachantus tayenus*) to reduce effort and catch, in order to maintain equilibrium as well as to avoid collapsed of this species in the fishery. The negative slope value showed that the population of this species was in critical point.

By delaying tc, from year 3 Threadfin big eye (*Priachantus tayenus*) had greater chances in recruitment. Nevertheless, this required further biological research to establish more in formation on Threadfin big eye (*Priachantus tayenus*).

A little amount of stock had increased in year 6 showing that this species had lower level of spawning stock. This would lead to viable recruitment and final alternative to enforce closure for certain period to ensure conservation target could be achieved.

The Threadfin big eye (*Priachantus tayenus*) fishery was biologically and economically under exploited. Looking at the historical data in Table 6.1. In 2001 the total effort was 1442 boats. This was the Maximum level in the range of optimum effort as determined through Surplus model. Therefore, the total effort in the fishery shouldn't be increased further in order to maximize the catch and profit. However, it is advisable to reduce the total mortality to 0.6 per year for the following reasons;

a. The efficiency of fishing gear is improving; therefore a total of 1442 boats at present might be equivalent to or less than 1000 boats in term of effort pressure in 50 years time.

b. Due to uncertainty of the environment in future, natural mortality might be increased.

c. These two parameters are assumed constant in this model. This study also indicate that the age at first captured should be maintained at 4 years old.

The collected catches in 1991 to 2000 as shown in Table 6.1. Using the Schaefer, the estimated MSY were 2594965.84 tonnes and 5333626.04 tonnes respectively. The estimated effort 946.891579 boats and 431.470361 boats respectively. These showed that the fishery were still biologically under exploited for next six years. However, the situation had to be reviewed after 6 years and precaution should be made earlier.

6.3.7 Catch

The result from all curve Age (year) against normal logarithms catch at tc equal tm (at year 6) had negative slope and this mean the population of this species was likely decreasing.

$$C = (F/Z) N (1 - \exp(-Z))$$

The population of Threadfin big eye (*Priachantus tayenus*) increased at 4th year, this mean there were recruitment in this population, where

$$R = aSm \exp(-bSm), \text{ Sm is a matured stock}$$

The result from curve Year (2003 to 2052) showed the level of population for this species decreased to critical point although there was recruitment but still could not recover caused by the high exploitation.

When combine, this curve produced negative slope to the surplus production model. The situation definitely showed that collapse had occurred in the fishery.

The performance of catch by changing mortality and decreased effort indicated that both numbers of mortality and catch were decreasing compared to total mortality of 0.823 per year. The reduction of total mortality had resulted good indicator to the fishery in long term for sustainable resources.

This model took into account recruitment, growth, mortality, fishing effort both in terms of number of boats and hours worked each boat per year. This includes the age at first capture, age at first recruitment, value of catch per kilogram. Assuming a constant mortality, the model could project the catch and profit for 50 years period. The model was designed for policy makers to review their policies.

This study also to determine the resource status of Threadfin big eye fisheries *(Priacanthus tayenus)* by using different parameter such as the mortality. This calculation was used for future management and suggestion for maintenance the sustainable stock. This model takes into account the recruitment, growth, mortality caused by fishing activities and natural mortality. Decreased through mortality either fishing mortality or natural mortality. Assuming an equal fishing effort, the model could project the catch and the profit for 50 years period.

A fish stock, subject to increasing presures, will diminish in size unless the rate of recruitment in the exploitation of fish stock is equal to or more than the total rates of exploitation and total mortality.

CONCLUSION AND RECOMMENDATIONS

7.1 Conclusions

Depending on ones orientation, Darvel Bay can be seen plays a variety of roles, a major fishing ground and aquaculture area of marine resources or a coastal tourism belt. Deforestation and utilizations of coastal forest and mangrove for industrial, urbanizations, shipping and port and infrastructure development have impact and damage to its marine environmental. There are also international pirates who operate and treat the vessel as well as local community. Conflicts among the uses and development in the bay arise at high level.

The survey area in Darvel Bay was done according to block, with consideration of trawl able fishing area 834.55 km square. The studies of demersal resources by using sweep area method. The result had shown that the potential resources were 1293723.93 kg and the density was estimated 1242.16 kg. or 1.2 metric tones per km square in this area.

There were 128 hauls with a total caught of 20,032.9 kg in this survey. Throughout species identification it was found 71 species of fish available. During this study the dominant species were Kerisi or Thereadfin bream (*Nemipterus spp*) 1001.645 kg and Duri/Pulutan/Utek or Marine catfish (*Tachsurus spp./Arius spp,Osteogenius spp*) 1001.645 kg or 5% followed by Batu or Parrot/Wrass (*Labridae scaridae*) 801.316 kg and Kerapu or Grouper (*Epinephelus spp*) 801.316 kg and Beliak mata/lolong bara or Thereadfin big eye (*Prianchantus tayenus*) 801.316 kg or 4%, Gelama/Tengkerong or Jewfish *(Sciaene spp./Johnius spp)* 600.987 kg, Kerisi bali or Sharptoothed bass *(Pristipormoides typus)* 600.987 kg, Kikek or Ponyfish (Slipmounth) or *Leiognathus spp.* 600.987 kg, chelek mata or blotched grunt (pomadasys spp.) and Selangat or Bony bream *(Anodontostoma chacunda)* 3%, Belanak Mullet *Liza spp./yalamugi spp* and Tenggiri or Spanish mackerel (*penaeid*) 3% respectively.

The ratio of catch composition of marketable fish was 7,606 kg or 37.96%, while underutilized fish was 12,426.9 kg or 62.03% and the average catch per hour was 39.12 kg.

Fisheries as a whole suffered from both commercial and traditional over fishing. The dilemma is that as the demands of fisheries resources increase the ability to sustain marine environmental may be decreasing. Thus, integrated coastal management and Ecological Sustainable Development are much needed to ensure the sustainability of resources. This is a challenge to the Fisheries Department of Sabah as well as fisheries manager to regulate exploitation prior to over harvest and must be consistent with the aims of maximizing management of the fish resources and minimizing logistical and technical limitation.

Fisheries sector are important and popular for economic development in every developed or developing country. This is because fisheries can produce food and income to the society in the coastal state. Most of these people depend income from fisheries. In other aspect, the fish stock are being exploited beyond Maximum Sustainable Yield (MSY) or over exploited and over capitalized where revenue can not cover the cost as a result the fishermen lost their profit. At this level the manager should include their program for recruitment to avoid depletion of fisheries stock in the area because the stock will not recover by recruitment of stock.

The resources value (RM000) for this species in 2003 was RM 333.98. After six modes through natural mortality and fishing mortality the total value RM 1593.43 In 50 years later or 2052 the biomass stock at first mode RM 3.0×10^{-7} and through recruitment, natural mortality (M) and Fishing mortality (F) the total value at six modes RM 1.6×10^{-6}

This phenomenon needed to be solved from becoming bigger and serious. The manager should protect the stock being over exploited by reducing by catch problem or reducing too many people depended on fisheries for their live hoods. This situation has led the fishing industries economically marginal, the stock being depleted and caused environmental damage because of fishing technologist and unhealthy fishing practice such as using dynamite and cyanide poisoning. In future global conservation standard should be restricted to all fisheries.

Base on input and output result, restriction could be imposed on *Prianchantus tayenus* fisheries by reducing effort and catch to maintain equilibrium on this species in the fishery. Besides, to avoid the species collapsed where the negative slope value had showed that population of the species is in critical point.

By delaying the age in first catch, from year 4 to year 5 *Prianchanthus tayenus* have greater chances to recruit. Only small increase occurred in year 6, this showed that this species had lower level for stock spawning that had led to viable recruitment. Final alternative to enforce closure on the fishery for a period of time to ensure conservation could be targeted.

The adoption of new technology by raising efficiency end profitability, results intensified fishing by established operators and also acts as an incentive for newcomers to adopt the new technology and enter the fishery. However, intensification tends to deplete natural resources which being a common property. Adversely affects all users that are exploiting them. Specifically stock depletion implies a fall in catch per unit of effort. Therefore, technological progress may contribute to the growth and development of fisheries but may also be indirectly responsible for over exploitation.

Gradual increase in the efficiency of fishing gear and methods sometimes referred to as 'technology creep' in a double-edged sword. If more fish are being caught for the same amount of apparent effort, profitability in the sort term will increase; but if catches continue to increase without control the fish stock will become over exploited.

The allocation of closing area to fishermen in that area, *Priachantus tayenus* fisheries as an incentive for fishermen to conserve the fish stock and at the same time to maximize their efficiency cost of harvesting. Closing area could be suitable for use in some other fisheries, so their success in improving the economic performance of *Priachantus tayenus* fishery is likely to influence the future direction of fisheries management in Sabah

No doubt the current policy is very hard to maintain due to large of fishing area and there are many small scale fishermen depend their live on this sector. Introducing the closing area mean the traditional fishermen have to find alternative job before the coming session. This policy would be difficult to be implemented because as usual no employment is available to unskilled workers in most developing countries. It is very hard to maintain the current policy in longer term where one hands grand money to contract new fishing vessel with other hand demolish old ones.

The over fishing had little to do with the fishermen who generate with highest profit in short time. Instant profit can be obtained with intensive method of catching. The big scale fisheries have big capital would leave the industries when the commodity collapsed compare to the traditional fishermen that fully depend on the commodity. Excess capacity and depletion of fishing stock will not always affect the big scale fishermen.

The price for fishing ground should be applied to the fishermen depends on the effort they use in this sector and price should be relevant to their income gained from this industries.

If the basic policy unchanged, permanent problem such as over capacity of the fleet, the processing industry, and over fishing will only accumulate. Societies need shift the burden of proof from demonstrating that on going or planned activities that will damage or destroy resources and adverse socioeconomic consequences to demonstrating that on going and proposed use will not reduce management option for 15-20 years for that reason.

The Maximum Sustainable Yields (MSY) estimated for Sabah *prianchantus tayenus* fishery with trawl indicated that the exploitation was currently over maximum level. However, the standard effort level slightly above the optimum effort level. Although the results obtained from the surplus production model analyses provides rough estimates, they are useful as preliminary information for the fisheries managers. However, further analyses are required using other more data-intensive techniques to provide accurate and precise assessments. By using HAFM model, the biomass for less or over 50 years can be forecast and predicted the status of number, biomass value of this species. This model also used for analyses recruitment, growth and mortality relationship. The recruitment increased the number of the stock while growth increased the weight while the number decreased because of fishing or natural mortality with this study forecast future management and suggestion to maintain the sustainable stock.

7.2　　　Recommendations

This study shows that:

a. Further study is needed to compare the results with the same species to be conducted in another locations such as Strait of Malacca or South China Sea.

b. In future, further research is needed to verify the present observations and results. The studies proposed to include another methodology. There is a possibility that further length of time be taken in this research to improve the result of this study.

c. In order to improve the catch of *Priachantus tayenus* it is necessary to catch the fish using highly constructed specific and selectivity nets to avoid error in the catch frequency and also to produce a good result. It is also important to catch the young and small fish.

d. It would be necessary in future that the research procedures should reduce the duration of time during net towing and use of large vessel with high power engine to catch more fish. Bigger space is needed to keep the catch fish to maintain their properties without any changes before weighing and measuring the length frequency.

e. At present the species are still not fully exploited. In future, the length of time to catch should be surveyed. Research needs to be conducted by bigger boat with high technology facility. It is necessary to help the management regime to manage these resources.

f. Moreover, a research project should be done on the impact of harvesting of this species on the other fish species, which depend largely on this species as example the predator species depend on the small species.

g. Further study would also be necessary to be verified if there are any changes in population of this species in this area due to the catch; and also to see whether there are any other species or the same species coming from other areas to replace the fish that have been harvested.

h. There are many types of nets used to catch *Priachantus tayenus*. Therefore, a study is needed to specify which catching method is suitable and efficient in terms of their selectivity; and also to reduce the by catch under-size animal from other species or the same species during harvesting with friendly environmental manner, since the Darvel Bay in Sabah is a suitable area for fish (*Priachantus tayenus*) breeding. This is shown by the occurrence of this species all the year.

i. A management regime needs to be set up to conserve this species in order to avoid harvesting during the spawning session or over- exploitation where recruitment cannot recover the parental catch due to the heavy harvesting.

Further research is also needed to study the history and parameter biology of this species and to introduce a closing season for harvesting as a conservation regime for this species.

j. The information of the fishery management objectives and policy, this study recognized the needs of information on the biological and the population dynamic of the Threadfin big eye *(Priacanthus tayenus)*, on the interest of the whole community and fishermen. The scientific research should be planned for long and short term management purposes with accountable living marine resources in the State.

REFERENCES/BIBLIOGRAPY

Alias, M. 1993(a). Mechanization of fishing boats and its consequences on the management of the fishery in the West Coast of Peninsular Malaysia. In proceeding of fisheries research seminar 1993, Terangganu. Department of Fisheries, Ministry of Agriculture, Malaysia.

Alias, M. 1993(b). Biological and economic yield from the multispcies fishery on the West coast of Peninsular Malaysia. Department of Fisheries. Unpublish.

Ambak, M.A. 1984. *Fish Communities in Paya Bungor, with Notes on its Development, Management and Recreational Use.* Ph. D. dissertation. Universiti Pertanian Malaysia.

Ambak, M.A. and Jalal, K.C.A., 1998. "Habitat Utilization by the Tropical Fish Community in the Man-made Lake Kenyir, Malaysia". *Fisheries Management and Ecology.* 5(2): 173-176.

Ambak, M.A., Jalal, K.C.A. and Jamaluddin, 1. 1998. "The Status of Recreational Fishes in a Tropical Rain Forest Reservoir, Malaysia". In *Malaysian Science and Technology Congress,* 7-9 November, 1998, Kuala Terengganu, Malaysia.

Anonymous. 1989. Deepsea fisheries resources within the Malaysian Exclusive Economic Zone. Seminar sumber perikanan laut dalam di zon economi eksklusif (ZEE) Malaysia. 20th Mei 1989. Kuala Lumpur.

Arms, K. and Camp, P.S. 1986. *Biology.* 3rd edn, Saunders College Pub, New York.

Arshad, A., B. Japar Sidik and M.S. Mohd. Zaki. 1997. Roles of mangrove ecosystem in Malaysia fisheries. Pages 95-104 in B. Japar Sidik, F.M. Yusof, M.S. Mohd. Zaki & T. Peter. eds., Fisheries and environment beyond 2000. Univertiti Putra Malaysia, Serdang, Malaysia.

Australia Bureau of Agriculture and Resources Economics. 1989. individual Transferable Quotas and The Southern BluefinTuna Fishery. Australia Government Publishing Service. Canberra.

Banster, K. and Camphalle, A. 1985. The Encyclopedia of Aquatic Life Facts on File. Inc. New York.

Baranov, T. I. 1918. On the question of the biological basis of fisheries of Nauch. Issledov. Iktiol. Inst. Izv., I, (1), 81-128 Moscow. (Rep. Div. Fish Management and Scientific Study of the Fishing Industry, I (1).

Baretta-Bekker, J.G., Duursma, E.K. and Kuipers, B.R. 1992. Encyclopedia of marine sciences. Springer Verlag, Berlin.

Berita Harian Sdn. Bhd. 2003. Program akuakultur terima peruntukan tambahan. Berita Harian 21 Ogos 2003: 3, 5.

Bell, F.W. 1978. Food from the Sea: the economics and politics of ocean fisheries. Westview Press, Boulder.

Bennet, I. 1992. Australian Sea-Shores. Angus & Roberson, New South Wales, Australia.

Berkes., Mahon.R., Conney. P.M., Pollnac. R., and Pomeroy. R. 2001. Managing small-scale fisheries alternative directions and methods International development research center. Otawa. 309pp.

Beverton, R.J.H. 1954. Notes on the use of theoretical models in the study of the dynamic of exploited fish population. U.S. Fish Lab., Beaufort, N.C., Misc. Contrib. 2: 159.

Beverton, R.J.H. 1962. Long term dynamics of certain North Sea Fish populations. Pp242-259 In: LeCren, E.D and M.W. Holdgate (eds), The Exploitation of natural animal population. Blackwell. Oxford.

Beverton, R.J.H. 1963. Maturation, Growth, and mortality of clupeid and Enggraulid stocks in relation to fishing. Rapp. P.-v. Reun. Cons. Perm. Int. Explor. Mer., 154: 44-67.

Beverton, R.J.H. and Holt. S. J. 1956. A review of methods for estimating mortality rate in fish populations with special reference to sources of bias in catch sampling. Rapp. P.-v. Reun. Cons. Perm. Int. Explor. Mer., 140: 67-83.

Beverton, R.J.H. and S. J. Holt. 1957. Notes on the dynamic of fish exploited fish populations. U.K. Min. Agric. Fish., Fisheries and Food. London. Invest. (Ser. 11) 19: 533.

Beverton, R.J.H. and Holt. S. J. 1959 (a). A Review of livespans and mortality rate of fish in nature and their relation to growth and other physiological characteristics. In CIBA found. Colloq. Ageing. V. The lifespan of animals. London: Churchill, pp 142-180.

Beverton, R.J.H. and Holt. S. J. 1959 (b). Manual of methods for fish stock assessment. Part ll. Table yield function, Rome: FAO Fish. Tech. Pap (38). 67 pp.

Bhattacharya, C.G. 1967. Simple method of resolution of a distribution into Gaussian components. Biometrics. 23: 115-135.

Birkes, D. and Dodge, Y. 1993. Alternative methods of regression. John Wiley and Sons. New York. 228pp.

Biusing, E.R. 1996. Status of the coastal fisheries resources of Sabah, Malaysia. Pages 31-80 in Chuan, T.T., Othman. M. and Fung, S., eds. Proceeding of the seminar on sustainable development of fisheries resources in Malaysia held at Kota Kinabalu, Sabah on 12-13 September 1995. Kota Kinabalu: Institute for development studies (Sabah).

Biusing, E.R. 1997. Investment opportunities in the marine fisheries sector of Sabah, Malaysia. Department of Fisheries – Sabah. Kota Kinabalu. 41pp.

Boaden, P. J. S., and Seed, R. 1985. An introduction to coastal ecology. Blackie, Glasgow. 218 pp.

Bond, C.E. 1979. *Biology of Fishes.* W.B. Saunders Company Philadelphia London Toronto.

Bond, C.E. 1996. *Biology of Fishes, Professor Emeritus of Fisheries and Wildlife* Oregon State University, Corvallis, Oregon.

Brown, D. and Rothery, P. 1993. Model in biology: mathematics, statistics, and computing. John Wiley and Sons. New York. 688 pp.

Buzeta, R.B. 1982. Report on the regional Training course on fishery stock assessment, 1 September - 9 October 1981, Samutprakarn, Thailand. Part II – Technical Report, Vol.1. 1982. South China Sea Fisheries Development and Coordinating Programme. Manila. 238 pp.

Carl, E.B. 1979. *Biology of Fishes.* Oregon State University, Corvallis, Oregon. 32 pp.

Castro, P. and Huber, M.E. 1991. Marine biology. Mosby Year Book, St.Louis.

Chang, L.W. 1977. Traditional Baited Bottom Long line (Rawai umpan) fisheries in Penang. Department of Fisheries, Ministry of Agriculture, Malaysia. Kuala Lumpur. 38 pp.

Chatterjee, S. and Price, B. 1977. Regression analysisi by example. John Wiley and Sons. New York. 228pp.

Chen, Y., Jackson, D.A., and Harvey, H.H. 1992. a comparison of von Bertalanffy and polynomial functions in modeling fish growth data. Canada Journal of Fisheries and Aquatic Science 49: 1228-1235

Chesson, J. The South East Fishery. 1995. Australian Fisheries Management Authourity. Canberra.

Chuan, T.T. Othman. M. and Fung, S.1997. Study on investment aspects of deep-sea fishing in Sabah. Institute for development studies (Sabah). Kota Kinabalu. 48pp.

Churchill, R.R. & Lowe, A.V. 1988. The law of the Sea, new rev. edn. Manchester University Press, Manchester.

CIA, The world factbook 2002. http://www.cia.gov/publications. Accessed on 27 June 2003

Clucas, I.J.1985. Fish handling, preservation and processing in the tropics: Part 2. Tropical Development and Research Institute. London. 143 pp.

Cochran, W.G. 1963. Sampling techniques 2nd ed. John Wiley and Sons, Ltd. London 413 p. New York. 228pp.

Coleman, N. 1991. Encyclopedia of Marine Animal. An Angus and Robertson Book, Australia. pp 32-34.

Coleman, N. 1997. A Field Guide To Australian Marine Life. Regby Limited, Australia.

Coleman, N. and Bennett, I. 1991. Encyclopedia of Marine Animals. An Angus and Roberson Book, Australia.
Coleman, N. and Seventy, V. 1987. Australian, Sea Life South of 30' S. Doubleday Australian Pty Limited, NSW, Australia.

Consumers Associaion of Penang. 1977. The Malaysian fisheries. The Consumers Association of Penang. Penang.

Couper, A. 1983. Atlas and Encyclopaedia of the sea. Times Books Limited, London.

Cunninham,S. Dunne, M. and Whitmersh, D. 1985. Fisheries economic: an introduction. Maxwell. London.

Cushing, D.H. 1984. Upelling and fish production. FAO Fish. Tech. Paper, (84): 40 pp.

Daim, B. 2001. Verbal conversation. Sabah Fisheries Department. Kota Kinabalu.

Dale, W.L. 1956. wind and drift currents in South East Asia. Malay. J. Trop. Georg. 8: 1-31.

Day, F. 1978. The fishes of India. Today and Tomorrow's book agency, New Delhi. 778 pp.

Day, J.H. 1967. A Monograph on the Polychaeta of Southern Africa. (Part I and II). Trustees of The British museum (Natural history). London. 842pp.

David, R. and George, J. 1996. Marine Life. Regby Limited, Australia.

Department of Fisheries. 1992. Traps. Department of Fisheries Sabah, Kota Kinabalu. 24 pp.

Department of Fisheries. 1999. Annual Fisheries Statistics. Department of Fisheries Malaysia, Ministry of Agriculture, Kuala Lumpur, Malaysia. pp. 42-43.

Department of Marine. 2002. Sabah – East Coast Darvel Bay, Map No. 1680. Taunton. United Kingdom.

Director of National Mapping, Malaysia. 1988. Published by the Director of National Mapping, Malaysia.

Doe, P.E. 1998. Fish Drying & Smoking. Technomic Publishing Co. Inc. Lancester. Pp 250.

Duncan, D.B. 1955. Multiple range and multiple F-tests. Biometrics. 11:1-42.

Eddie, G. 1980. Technology and fisheries development. London Society for under water technology. London.

Edgar, G.J. 1997. Australian Marine Life. Reed Book, Australia.

Edwin, S.I. 1996. Living Marine Resources Their Utilization and Management. Chapman & Hall. New York.

FAO. 1990. World statistic yearbook. Rome.

FAO. 1991. World statistic yearbook. Rome.

FAO. 1992. World statistic yearbook. Rome.

FAO. 1993. World statistic yearbook. Rome.

FAO. 1994. World statistic yearbook. Rome.

FAO. 1995. World statistic yearbook. Rome.

FAO, 1982 Fish by catch…Bonus from the sea. Rome.

FAO. 1985. Small-scale processing of fish. International labour office Switzerland.

FAO. 1997. Review of the state of world fishery resources: marine Fisheries. Rome.

FAO. 1987. Manual for the Management of Small Fishery Enterprises. Rome.

FAO. 1988. Contribution to tropical fisheries biology. Rome.

Fishbase. 2003. *Priacanthus tayenus*
    http://www.fishbase.org. Accessed on 20 June 2003

Fisheries Department, Sabah. Annual Report. 1975. Kota Kinabalu.

Fisheries Department, Sabah. Annual Report. 1990. Kota Kinabalu.
    Fisheries Department, Sabah. Annual Report. 1992. Kota Kinabalu.

Fisheries Department, Sabah. Annual Report. 1993. Kota Kinabalu.

Fisheries Department, Sabah. Annual Report. 1994. Kota Kinabalu.

Fisheries Department, Sabah. Annual Report. 1995. Kota Kinabalu.

Fisheries Department, Sabah. Annual Report. 1996. Kota Kinabalu.

Fisheries Department, Sabah. Annual Report. 1997. Kota Kinabalu.

Fisheries Department, Sabah. Annual Report. 1998. Kota Kinabalu.

Fisheries Department, Sabah. Annual Report. 1999. Kota Kinabalu.
    2000. Kota Kinabalu.

Ford, E., 1933. Population age structure in males and juveniles of the Antartic krill, *Euphasia superba*. Dana, Polar Biol. 4: 199-201.

Fox, W.W. 1970. An exponential surplus-yield model for optimizing exploited fish populations. Transactions of the American Fish Society 99: 80-88.

Fox, W.W. 1975. Fitting the generalized stock production model by least-squares and equilibrium approximation. Fishery bulletin 73: 23-37.

Fudd, S. 1971. Inshore Fishing. Fishing News (books) Ltd, London. 131 pp.

Gardon, H.S. 1954. The economic theory of a common property resources: the fishery. Journal of political economy, LXII(2), 124-42.

Gaskel, T.F. 1970. World beneath the oceans. International Learning Systems Corporation Limited. London. 154 pp.

Gayanilo, F.C.; P. Sparre and D. Pauly. 1996. FisAT: FAO – ICLARM STOCK ASSESSMENT TOOLS, User's manual, Rome, FAO.

Gayanilo, F.C.; P. Sparre and D. Pauly. 2002. FisAT 11: User's manual, Rome, FAO.

Gulland, J.A,. 1955(a). Estimation of growth and mortality in commercial fish population. U.K. Min. Agric. Fish., Fish. Invest. (Ser. 2) 18(9): 46.

Gulland, J.A,. 1955(b). On the estimation of population parameters from mark ed numbers. Biometrika 42: 269-270.

Gulland, J.A,. 1956. On the fishing effort in English demersal fisheries. Fishery Inves. Ser(ll) 20: 1-41.

Gulland, J.A,. 1965. Estimation of mortality rate, ICES CM, 3, 1.

Gulland, J.A,. 1969. Manual of methods for for fish stock assessment. Part 1. Fish population analysis. FAO Man. Fish. Sci. 4; 1-154.

Gulland, J.A. 1971(a). Manual of Methods for fish stock assessment, Part 1. Fish Population Analysis. 1969. Food and Agriculture Organization. Rome. 154 pp.

Gulland, J.A. 1971(b). The Fish Resources of the Ocean. Surrey: FAO/Fishing News (Book) Ltd. London.

Gulland, J.A. 1974. Guidelines for fisheries management. Rome: FAO.

Gulland, J.A,. 1983. Fish stock assessment: A manual of basic methods. Rome: FAO-john Wiley and Sons. 223 pp.

Gulland, J.A,. 1988. Fish population Dynamics: the implications for management. John Wiley and Sons. Chichester.

Gulland, J.A,. and Holt, S.J. 1959. Estimation of growth parameters for data at unequal time intervals. J.Cons. Int. Explor. Mer. 25(1): 47-49.

Gulland, J.A,. Rosenberg, A.A,. 1992. A review of length-based approaches to assessing fish stocks. Food and Agriculture Organization. Rome. 100 pp.

Gulland, J.A,. Rothschild, B.J.1984. Penaeid shrimps – their biology and management. Fishing News (books) Ltd, London. 131 pp.

Graham, M. 1935. Modern theory of exploiting a fishery and application to North Sea trawling. J. Cons. Int. Explor. Mer, 10, 264-274.

Grove, D., Cord and Hunt, L.M. 1980. The Ocean World Encyclopaedia. Mc Graw-Hill Book Company, Australia.

Haddon, M. 2001. Modelling and quantitative methods in fisheries. Chapman & Hall/CRC. Florida. 406 pp.

Hassal and Associated Pty Ltd. 1988. Underutilised seafood species and export market, Report for the Australian Fisheries Service. Canberra.

Hilborn, R. 1979. Comparison of fisheries control systems that utilize catch and effort data. Journal of the fisheries Research Board of Canada 36: 1477-1489.

Hilborn, R., Sibert, J. 1988. Adaptive management of developing fisheries, Marine Policy April, 112-21.

Hilborn, R., and Walters, C.J. 1992. Quantitative Fisheries Stock Assesment. Chapman & Hall, Inc. London. 570 pp.

Hon, H. 1992. Ecology vs economy. Science and Technology Malaysia, August: 2, 6-12.

Ingles, J. and Pauly, D. 1984. An atlas of the growth, mortality, and recruitment of the Philippenes Fishes. ICLARM Tech. Pap.(13): 1-127.

Jabatan Kajicuaca Malaysia. 1998. Ringkasan Bulanan Pemerhatian Kajicuaca tahun 1998.

Jamaluddin, MD. J. 1989. Pengantar Geomorfologi. Dewan Bahasa dan Pustaka. Kementerian Pendidikan Malaysia, Kuala Lumpur. 404 pp.

Jennings, S. Kaiser, M. Reynolds, J. 2001. Marine Fisheries Ecology.
 Blackwell Science. United Kingdom.

Jomo, K.S. 1991. Fishing for Trouble. Institute for Advanced Studies, University of Malaya. Kuala Lumpur.

JUPEM. 2000. Jadual ramalan air pasang surut (Tide Tables) Malaysia. Diterbitkan oleh Jabatan Ukur dan Pemetaan Malaysia.

Kailola, J., Williams, M.J., Stewart, P.C., Reichett R.E., McFee,A. and Grieve, C. 1993. Australian Fisheries Resources. Bureau of Resource Sciences and Fisheries Research and Development Corporation, Canberra, Australia.

Kennish, M. J. 1994. Practical handbook of marine science (2nd ed.). Crc press, Florida. 566 pp.

Kershaw, D.R. 1982. Animal Diversity. Angus & Roberson, New South Wales, Australia.
Kimura, D.K. 1980. Likelihood methods for the von Bertalanffy growth curve. U.S Fishery Bulletin 77: 765-776.

King, M. 1996. Fisheries Biology, Assessment and Management. Fishing News Books. Great Britain. 341pp.

Kish, L. 1987. statistical design for research. John Wiley and Sons, Inc. London. Pp 1-26.

Kohler, H. 1988. Statistics for Business and economic. Scott, Foresman and Company Glenview, Illinois London, England.

Kristjonsson, H. 1975. Modern Fishing Gear of the World. . Food and Agriculture Organization. Rome. 607 pp.

Lablache, G. and Carrara, G. 1988. Population dynamics of Emperor Red Snapper (Lutjanus sebae), with notes on the Demersal fishery on the Mahe' Plateau, Seychelles. Pages 171-192 in Venema, S.C., Christensen, J. M., and Pauly, D. eds. Contributions to tropical fisheries biology. Papers prepared by the participants at the FAO/DANIDA Follow-up Training Courses on fish stock assessment in the tropics. Hirtshals, Denmark, 5-30 May 1986 and Manila, Philippines, 12 January-6 February 1987. FAO Fish.Rep., (389):519pp.

Lalli, C. and Parsons, T.R. 1993. Biological oceanography. Pergamon Press, Oxford. U.K.

Leslie, P.H. 1945. The use of matrices in certain population mathmetics. Biometrika, 35, 213-45.

Levinton, J.S. 1995. Marine Biology, Function, Bio diversity and Ecology. Oxford University Press, New York.

Lobeck, A. K. 1981. Geomorphologi. Dewan Bahasa dan Pustaka Kementerian Pelajaran Malaysia, Kuala Lumpur. 920 pp.

Ludwig, D., and Walters, C.J. 1981. Measurement errors and uncertainty in parameter estimation for stock and recruitment. Canadian Journal of fisheries and Aquatic Science 38: 711-720.

Ludwig, D., and Walters, C.J. 1985. Are age structured models appropriate for catch-effort data? Canadian Journal of fisheries and Aquatic Science 42: 1066-1072.

Ludwig, D., and Walters, C.J. 1988. Comparison of two models and two estimation methods for catch and effort data. Natural Resources Modelling 2: 457-498.

Malaysia Ministry of agriculture. 1990. Annual fisheries statistics. Kuala Lumpur.

Malaysia Ministry of agriculture. 1991. Annual fisheries statistics. Kuala Lumpur.

Malaysia Ministry of agriculture. 1992. Annual fisheries statistics. Kuala Lumpur.

Malaysia Ministry of agriculture. 1993. Annual fisheries statistics. Kuala Lumpur.

Malaysia Ministry of agriculture. 1994. Annual fisheries statistics. Kuala Lumpur.
Malaysia Ministry of agriculture. 1995. Annual fisheries statistics. Kuala Lumpur.

Malaysia Ministry of agriculture. 1996. Annual fisheries statistics. Kuala Lumpur.

Malaysia Ministry of agriculture. 1997. Annual fisheries statistics. Kuala Lumpur.

Malaysia Ministry of agriculture. 1998. Annual fisheries statistics. Kuala Lumpur.

Malaysia Ministry of agriculture. 1999. Annual fisheries statistics. Kuala Lumpur.

Malaysia Ministry of agriculture. 2000. Annual fisheries statistics. Kuala Lumpur.

Marr, J.C. 1951. "On the use of term abundance, availability, and apparent abundance in fishery biology". Copeia. 1951;163-169.

Marzillier, L.F. 1990. Elementary Statistics. Wm.C.Brown Publishers, Dubuque.

Maxwell, C.N. 1921. Malayan fishes. Methodist Publishing House. Singapore. 280 pp.

Mayer. A.G. 1912. Ctenophores of the Atlantic Coast of North America. Canegia Institute, Washington D.C.

McBride, E. F. 1971. Mathematical treatments of size distribution data. Pages 116-121 in Carver, R. E. (ed.) Procedures in sedimentary petrology. Wiley Interscience, New York.

McConnaughey, B.H. and Zottoli, R. 1983. Introduction to marine biology. 4th edn, Mosby, St. Louis.

Meadows, P.S. and Campbell, J.I. 1988. An introduction to marine science. 2nd edn, Blackie, Glasgow.

Meddis, R. 1975. Statistical handbook for non-statisticians. McGraw-Hill Book company (UK) limit. London.

Megrey, B.A. 1989. Review and comparison of age-structured assessment models from a theoretical and applied point of view. American Fisheries Society symposium, 6, 8-48.

Mohammad Shaari, S.A.L., Weber, W., Kean, L.A., and Chang, L.W. 1976. Demersal fish resources in Malaysian water-6. Ministry of Agriculture and Rural Development, Malaysia. 64 pp.

Mohd. Noh, K. *et al.* 1992. Fish Industry: Prospects and Challenges. Malaysia Fisheries Society Occasional Publication No. 5. Kuala Lumpur.

Mohsin, A.K.M., and Ambak, M.A. 1996. Marine fishes and fisheries of Malaysia and neighbouring countries. Universiti Pertanian Malaysia, Serdang, Malaysia. 744pp.

Mohsin, A.K.M., and Ambak, M.A., Salam, M.N.A. 1993. Malay, English and scientific names of fishes of Malaysia. Universiti Pertanian Malaysia, Serdang, Malaysia. 227pp.

Morse, J.L. Hendelson, W. H. 1971. (a). Funk & Wagnalls New Encyclopedia Vol.6 Funk & Wagnall's, Inc. New York.

Moyle.P.B., and Cech. J.J, Jr. 2000. Fishes an Introduction to Ichthyology. Prentice-Hall, Inc. New Jersey. 612 pp.

Moore, G., and Jennings, S. 2000. British Ecological Society. Great Cem, P.S. 1989 Bratain. Pp 66.

Munro, J.L., Caribbean coral reef fishery resources. International Center for Living Aquatic Resources Management. Manila. 276 Pp.

Nelson, J.S,. 1994. Fishes of the world. John Wiley & Sons, Inc. New York. Pp 471

Nicholas, V.C., Polunin and Roberts, C.M. 1996. Reef fisheries. Chapman & Hall. London.

Nikolsky, G.V. 1963. *The ecology of Fishes*. Academic Press, London and New York. Pp.295 pp.

Nedelec, C. 1987. Catalogue of Small Scale Fishing Gear. Food and Agriculture Organization. Rome. 224 pp.

Parker, S. (ed.) 1980. McGraw-Hill encyclopedia of ocean and atmospheric sciences. McGraw-Hill, New York.

Pauly, D,. 1978. A preliminary compilation of fish length growth parameter. Ber. Inst. F. Meereskunde (kiel University), 55: 200.

Pauly, D,. 1979(a). Gill size and temperature as governing factors in fish growth: A generalization of von Bertalanffy's growth formula. Ph.D. Thesis. Institut-fiir Merreskunde. An. Der. Christian - Albrechts – Universitat Kiel, 155pp.

Pauly, D,. 1979(b). Theory and management of tropical multispecies stock: A review, with emphasis on the Southeast Asian Demersal fisheries. ICLARM stud. Rev. 1, 35pp.

Pauly, D,. 1980(a). On the interrelationships between natural mortality, growth parameter, and mean environmental temperature in 175 fish stock. J.Cons.CIEM, 39(2):175-192.

Pauly, D,. 1980(b). A selection of simple methods for the assessments of tropical fish stocks. FAO Fish. Circ. 729: 54.

Pauly, D,. 1981(a). The relationships between gill surface area and growth performance in fish: a generalization of von Bertalanffy's theory of growth. Meeresforsch 28: 251 – 282.

Pauly, D,. 1981(b). Some simple method for the assessment of tropical fish stocks. FAO Fish. Tech. Pap. 234(viii): 52.

Pauly, D,. 1982. The fishes and their ecology. In Pauly. D. and Mines, A.N. (eds) Small-scale fisheries of San Miquel Bay, Philippines: Biology and stock assessment, Manila: ICLARM Tech. Rep. &, pp15-33.

Pauly, D,.1983(a). Theory and Management of Tropical Multispecies Stock International Centre for Living Aquatic Resources Management Manila, Philippine. Manila.

Pauly, D,.1983(b). Length-converted catch curves. A powerful tool for fisheries research in the tropics. (Part I). ICLARM Fisbyte, 1(2):9-13.

Pauly, D,. 1984(a). Fish population dynamics in tropical waters: a manual for use with programmable calculators. ICLARM Stud. Rev. 8,325pp.

Pauly, D,. 1984(b). Length converted catch curve: A powerful tool for fisheries research in tropics (Part ll). ICLARM Fisbyte, 1(2):17-19.

Pauly, D. and David. N. 1980. A basic program for the objective extraction of growth parameters from length frequency data. Inter. Counc. Explor. Sea C.M. 1980/D7. Statistics Committee, p. 33.

Pauly, D. and Gaschultz, G. 1979. Simple method for fitting on e: A powerful tool for fisheries research in tropics (Part lll). ICLARM Fisbyte, 2(3):9-10. Oscillating length growth data, with a program for pocket calculators. ICES: CM 1979/G.

Pauly, D. and Ingles. J,. 1981. Aspects of the growth and natural mortality of exploited coral reef fishes. Paper presented at the 4th international Coral reef Symposium, 17 – 25 May, 1981. Manila , Philippines, p. 24.

Pauly, D. and Mines, A.N. 1982. Small-scale fisheries of San Miguel Bay, Philippines biology and stock assessment. Institute of Fisheries Development and Research Collegge of Fisheries, University of Philippines, Visayas, Quezon City, 123 pp.

Pauly, D., Moreau, J. and Abad, N. 1995. Comparison of age structured and length converted catch curves of brown trout salmo trutta in two French rivers. Fisheries Research 22, 197-204.

Pauly, D. and Morgan, G.R. 1987. Length-based methods in fisheries research.International Centre. Manila. 468 pp.

Pauly, D. and Munro, J. 1984. One more on the comparison of growth in fishes and invertebrates. Fishbyte. 2(1):21.

Pauly, D. and Soriano, M.L 1986. some practical extensions to Beverton and Holt's relative yield-per-recruit model. In MacLean, J.L., Dizon, L.B. and Hosillos (eds). The first Asian fisheries forum. Mannila: Asian Fisheries Society, pp. 491-495.

Petersen, C.G., 1891. Eine Methode zur Bestimmung des Alters und Wuchses der Fishe. Mitteilungen. Deutscher Seefisherei-Verein 11: 226-235.

Pilcher, N., and Cabanban, A. 2003. Status: Coral reefs
http://www.reefbase.org   Accessed on 17 July 2003

Pitcher, T.J., and Hart, P.J.B. 1982. Fisheries ecology. Croom Helm. London. 414 pp.

Powel, A.B. 2003. *Priacanthidae*. http://www4.cookman.edu/noaa/ichthyoplankton/priacanthidae. Accessed on 27 November 2003.

Powell, D.G. 1979. Estimation of growth and mortality parameters from the   length-frequency of catch. Rapp. P.-v. Reun. Cons. Int. Explor. Mer. 177: 466-476.

Rahmat, L. 2001. Verbal conversation. Sabah Fisheries Department. Kota Kinabalu.

Rees, D. G. 1995. *Essential Statistics.* Third Edition, Chapman and Hall, 147 pp.

Ricker, W.E,. 1975. Computation and interpretation of biological statistic of fish populations, Bull. Fish. Res. Board. Can. (191).

Riesenfeld, S.A. 1971. Protection of Coastal Fisheries Under International Law. Johnson Reprint Corporation. London. 296 pp.

Sabli, M. 2001. Verbal conversation. Sabah Fisheries Department. Kota Kinabalu.

Schaefer, M.B. 1954. Some aspects of dynamics of populations important to the management of commercial marine fisheries, Bull. I-ATTC/Bol.CIAT, 2:247-68.

Schaefer, M.B. 1954. Some aspects of dynamics of populations important to the management of commercial marine fisheries, Inter-American Tropical Tuna Commission Bulletin, 1(2) 27-56.

Schaefer, M.B. 1957. A study of the dynamics of the fishery for yelloefin tuna in the Eastern Tropical Pacific Ocean. Bulletin, Inter-American Tropical Tuna Commission 2: 247-284.

Scott, J.S. 1959. An introduction to the sea ffishes of Malaya. Ministry of Agriculture Malaya. Kuala Lumpur. 179 pp.

Sikorski, Z.E. 1990. Seafood: Resources, Nutritional Composition, and Preservation. CRC Press, Inc. Florida.

Silvestre, T. G. and Pauly., D,. 1987. Estimate of yield and economic rent from Philippine Demersal stocks (1946-1984) using vessel horsepower as an index of fishing effort. Univ. Philipp. Visayas Fish. J. 1-3, 11-24.

Smith, T.D. 1988. Stock assessment methods: The first fifty years. In: Fish population dynamics: The Implications for management. (ed) Gulland, J.A. 1-33. John Wiley and Sons, United Kingdom. 422 pp.

Somo, K, Abu Bakar, S, and Shahardin, Z.A. 1990. Report on the nautical data of Matahari Expedition 1989, Edited by Abu Khair Mohammad Mohsin, Mohd. Zaki Mohd. Said and Mohd. Ibrahim Hj.Mohamad, Faculty of Fisheries and Marine Science, Universiti Pertanian Malaysia, Occasional Publication No.9. Malaysia.

Part 1 – Manual. Food and Agriculture Organization. Rome. 337 pp.

Sparre, P. and Siebseh. 1992. Introduction to tropical fish stock assessment. Part I Manual. FAO, Rome. 375 pp.

Tawar, P. K., and Kacker, K. K. 1984. Commercial sea fishes of India. The Pooram, Calcuta, India.

Taylor, C.C. 1958. Cod growth and temperature. J. Cons. Inter. Explor. Mer. 23, 366-370.

Tony. J, P. 1982. Fisheries ecology. The Avi Publishing Company inc.Connecticut.

Unnip, T. "Sejuta hektar karang musnah" Berita Harian, 6th August 2001, 6, 1-4.

Veterinary Department Sabah. 2002. Annually Report 2001. Kota Kinabalu.

von Bertalanffy, L. 1938. A quantitative theory of organic growth. Human Biology 10:181-213.

Walford, L.A. 1946. A new graphic method of describing the growth of animals. Biol. Bull. 90(2): 141-147.

Waugh, G. 1984. Fisheries Management Theoretical Developments and Contemporary Applications. West view Press/Boulder, Colorado.

Wetherall, J.A. 1986. A new method for estimating growth and mortality parameters from length frequency data Fishbyte 4, 12-14

Wetherall, J.A., Polovina, J.J. and Ralston, S. 1987. Estimating growth and mortality in steady state fish stocks from length-frequency data. In the ICLARM/KISR
Conference and the theory and application of length –based method for stock assessment, 11-16 February, Sicily, Italy.

Younger, M.S. 1979. Handbook for linear regression. Duxberry Press. Belmont. 569 pp.

APPENDIX

Appendix 1: Fishes survey form: HA/kustem/cap7/1

Date…………………………….

Location:………………………….

Species……………………………

Fishing Gear………………………

Mesh Size:………………………..

No	Size(SL)	Weight (Gram)

Appendix 2: Fishes survey form: HA/kustem/cap5/1

Date……………………….

Location:…………………..

Fishing Gear…………………

Mesh Size:……………….

Group Categories of Fishes: **2 Diadromous Fishes**

ISSCAAP Code	Division Group of Species	Local Name	English Name	Scientific Name	Weight (Kilogram)
24	Shad, Milk-Fishes, Etc.	Kebasi/ Selangat	Chacunda shad (Gizzard shad)	Anodontosoma	
		Puput	Shad	Pellona spp.	
		Beliak mata	Slender shad	Shadllisha elongata	
		Terubok	Longtail shad	Shad hilsa macrura	
25	Miscellaneous	Siakap	Giant sea perch (Barramudi)	Lates calcarifer	

HAMID AWONG FISHERIES MODEL (HAFM)

Appendix 3: Fishes survey form: HA/kustem/cap5/2

Date……………………….

Location:…………………..

Fishing Gear…………………

Mesh Size:…………………..

Group Categories of Fishes: **2 Marine Fishes**

ISSCAAP Code	Division Group of Species	Local Name	English Name	Scientific Name	Weight (Kilogram)
31	Flounders, Halibut, Soles, Etc	Lidah	Tonguefish (Tongue sole)	*Cynoglossus*	
		Sebelah	Flatfish	Pseudorhombus spp.	
33	Redfishes, Basses, Congers, Etc.	Bayan	Parrotfish	Callyodon spp. Thalassoma spp.	
		Jarang gigi	Croacker	*Otolithus*	
		Batu	Parrot/Wrass	Labridae scaridae	
		Biji nangka	Groatfish	Upeneus spp.	

ISSCAAP Code	Division Group of Species	Local Name	English Name	Scientific Name	Weight (Kilogram)

ISSCAAP Code	Division Group of Species	Local Name	English Name	Scientific Name	Weight (Kilogram)
		Daun baharu	Spotted sicklefish	*Drepane punctata*	
		Delah/sulit	Fussiller	*Caesioerythrogaster/ C.chrysona*	
		Dengkis/Debam	Spinefeet (Rabbitfish)	*Siganus spp.*	
		Duri/Pulutan/Utek	Marine catfish	*Tachsurus spp./Arius spp, Osteogenius spp.*	
		Gelama/Tengkerong	Jewfish	*Sciaene spp./Johnius spp. Osteogenius spp.*	
		Gerut-gerut	Grunter	*Promadasys spp.*	
		Jebong	Triggerfish	*Abalistes stellaris*	
		Jenahak	Mangrove snapper	*Lutianus johni*	
		Kaci	Sweetlip	*Spilotichytyhs picfus*	
		Kapas Laut	Majorras (Silver bides)	*Gerres filamentosus G. abreviatus*	
		Kerapu	Grouper	*Epinephelus spp. Plectropornus spp.*	
		Kerisi	Thereadfin bream	*Nemipterus spp*	

ISSCAAP Code	Division Group of Species	Local Name	English Name	Scientific Name	Weight (Kilogram)
		Kerisi bali	Sharptoothed bass	*Pristipormoides typus*	
		Kikek	Ponyfish (Slipmounth)	*Leiognathus spp. Gazz spp, Secutor spp.*	
		Lumi-lumi	Bombay-duck	*Harpodon nehereus*	
		Malong	Conger eel	*Muraenesox spp.*	
		Merah	Red snapper	*Lutianus argentimaculatus/ L.malabaricus*	
		Mengkerong/Ubi/ Conor/Gadong	Lizard fish	*Sauride spp.*	
		Puntung Damar/ Bulus-bulus	Sillago-whitings	*Silago sihama/ s.maculuta*	
		Pasir-pasir/Timun-Timun/ Puyu Laut	Monocle bream	*Scolopsis spp.*	
		Pelandok	Emperors (Scavengers)	*Lethrinus spp.*	
		Pluru	Spadefish	*Ephippus orbis*	

		Remong/Kunyit-kunyit	Snapper	*Lutianus vitta/ L.lineolatus*
		Semilang	Catfish eel	*Plotosus spp.*
		Shrumbu/Lemah	False trevally	*Lactarius lactarius*
		Tanda	Snapper	*Lutianus russelli*
34	**Jacks, Mullets, Sauries, Etc.**	Aji-aji	Amberjack	*Seriolanigrafasciata*
		Alu-alu/Kacang-kacang	Barracuda	*Syhyraena jello/S.optusa*
		Aruan tasek	Black kingfish	*Rachycentron canadus*
		Bawal hitam	Black pomfret	*Formio niger*

ISSCAAP Code	Division Group of Species	Local Name	English Name	Scientific Name	Weight (Kilogram)
		Bawal putih	Whhite pomfret	*Pampus argenteus*	
		Bawal tambak	Chinese pomfret	*Pampus chinensis*	
		Bawal selatan	Small pomfret	*Pampus spp.*	

		Belanak/Kedera	Mullet	*Liza spp./yalamugi spp.*	
		Cermin/Sagai/Cupak	Horse mackerel (Trevally)	*Alectis indica/Ccaranx spp.*	
		Cincaru	Hardtail scad	*Megalaspis cordyla*	
		Demudok/Rambai	Horse mackerel (Trevally)	*Carangoides spp*	
		Gerong-gerong	Golden trevally	*Caranx speciosus*	
		Kurau/Senangin/Senohong	Thereadfin	*Polynemus spp./ Eleutheronen Tetradactylum*	
		Kerepoh	Horse mackerel	*Caranx sexfasciatus*	
ISSCAAP Code	Division Group of Species	Local Name	English Name	Scientific Name	Weight (Kilogram)
		Lolong	Ox-eye scad	*Selar boops*	
		Mata besar/Selar	Big eye scad	*Selar crymenophthalmus*	
		Pisang-pisang	Rainbow runner	*Elagatis bipinnulatus*	

		Selar/Pelata/ Temanong	Selar scad	*Selar spp.*	
		Selar kuning	Yellow Striped (Travally)	*Selaroides leptolepis*	
		Selayang/Curut	Round scad	*Decapterus maruadis/ D.macrosoma*	
		Talang	Queenfish/ Leatherskin	*Scomberoides commersonianus*	

ISSCAAP Code	Division Group of Species	Local Name	English Name	Scientific Name	Weight (Kilogram)
35	**Herrings, Sardines, Anchovies, Etc.**	Tamban sisik	Fringescale	*Sardinella fimbriate*	
		Bilis	Anchovy	*Stolephorus spp.*	
		Parang-parang	Dorab wolf-herring	*Chirocentrus dorad*	
		Bulan-bulan	Indo-Pacific torpon	*Megalops cyprinoides*	
36	**Tunas, Bonitos**	Aya/Kayu/Tongkol hitam	Longtail tuna	*Thunnus tonggol*	
		Aya/Kayu/Tongkol kurik	Kawa-kawa (eastern little tuna)	*Euthynnus affinis*	

	Billfishes, Etc	Aya/Kayu/Tongkol selasih	Frigate tuna	*Auxis thazard*	
		Layaran/Mersuji	Sailfish/Black & Blue marine	*Istiophorus spp./Makira spp.*	
		Tenggiri	Spanish mackerel	*Scomberomorus spp.*	

ISSCAAP Code	Division Group of Species	Local Name	English Name	Scientific Name	Weight (Kilogram)
37	Mackerels, Snoeks, Cutlass, Fishes, Etc.	Kembong /Rumahan	Indian mackerel	*Rastrelligerka nagurta/R. Fauhni*	
		Timah/ Layor/selayor	Large-head hairtail (Ribbon fish)	*Trichiurus lepturus*	
		Tulai	Japanese mackerel	*Scomber australasicus*	
38	Sharks, Rays, Chimaeras, Etc.	Yu	Shark	***Galeorhinidae***	
		Pari	Ray	*Gymnura spp. Dasyatis spp.*	
39	Miscellaneous marine fishes	Ikan Baja	Trash fish	*Mixed spp.*	
		Ikan campor	Mixed fishes	*Mixed spp.*	

Appendix 4: Fishes survey form: HA/kustem/cap5/2

Survey Form: HA/kustem/cap5/2

Date……………………………..

Location:……………………..

Fishing Gear…………………

Mesh Size:……………………

Group Categories of Fishes: **4 CRUSTACEANS**

ISSCAAP Code	Division Group of Species	Local Name	English Name	Scientific Name	Weight (Kilogram)
42	Sea spiders, Crabs. Etc.	Ketam laut/Ketam Renjong		*Purtunus pelagicicus*	
		Ketam batu/bakau Renjong	Mud crab	*Scyllaserrata*	
		Berangkas	Horse shoe crab	*Paralithodes spp.*	
43	Lobsters, Spinny rock Lobster, Etc.	Udang karang	Spiny lobster	*Panulirus polyhagus*	
		Udang lobak	Slipper Lobster	*Thenus orientalis*	

HAMID AWONG FISHERIES MODEL (HAFM)

ISSCAAP Code	Division Group of Species	Local Name	English Name	Scientific Name	Weight (Kilogram)
45	Shrimps, Prawn, Etc.	Bubuk / udang baring	Small shrimps	*Labridae scaridae*	
		Udang putih / udang kaki	Banana prawns/Western	*Penaeus merguiensis / P.Indicus/ P. Latisulcatus*	
		Udang harimau	Giant tiger prawn/Green Tiger prawn	*Penaeus monodon / P. Semisulcatus*	
		Udang Pasir/kepala Besar	Sand prawn	*Metapeneopsis stridulans & M.berbeensis/ Trachpenaeus fulvus*	
		Udang putih kecil	Small white prawn	*Metapenaeus lysianassa*	
		Udang kuning	Yellow prawn	*Metapenaeus brevicornis*	
		Udang merah ros / U.ekor biru	Pink prawn / Greasy back prawn	*Metapenaeus Affinis/M. Ensis/M. Intermedius*	
		Udang kulit keras	Rainbow prawn	*Parapenaeopsis sculptilis*	

ISSCAAP Code	Division Group of Species	Local Name	English Name	Scientific Name	Weight (Kilogram)
		Udang minyak/U. Minyak jalur	Sharp-rostrum prawn	*Parapenaeopsis Hardwickii/P. coromandelica*	
		Udang meerah	Red prawn	*Solenocera Subnuda*	
		Lain jenis Udang	Other prawns		

HAMID AWONG FISHERIES MODEL (HAFM)

Appendix 5: Fishes survey form: HA/kustem/cap5/3

Survey Form: HA/kustem/cap5/3

Date……………………………

Location:………………………

Fishing Gear…………………

Mesh Size:……………………

Group Categories of Fishes: **5 MOLLUSCS**

ISSCAAP Code	Division Group of Species	Local Name	English Name	Scientific Name	Weight (Kilogram)
53	**Oysters**	Tiram	Rock oysters/ Flat	*Oyster Ostea folium/ Crassostrea spp.*	
54	Mussels	Kupang/ Siput sudu/ /	Green mussel	*Pernaviridis*	
		Siput cangkul	Sea green mussel	*Glauconome spp.*	
		Siput mentiah	Abalone	*Haliotis spp.*	
56	Clams, Cockles, Arkshells, Etc.	Kerang	Blood cockle	*Andara granosa*	

ISSCAAP Code	Division Group of	Local Name	English Name	Scientific Name	Weight (Kilogram)

	Species			
		Retak serbu	Carpet clam	*Paphia undulata*
		Lain-lain siput	Other clams/snails	*Bivalves/Gastropods*
57	**Squids, Cuttlefishes, Octopuses, Etc.**	Sotong biasa/cumit-cumit	Common squids	*Loligo spp.*
		Sotong katak	Cuttlefish	*Sephia spp.*
		Sotong kereta	Octopus	*Octopodidae*

Appendix 6: Fishes survey form: HA/kustem/cap5/4

Survey Form: HA/kustem/cap5/4

Date.................................

Location:...........................

Fishing Gear......................

Mesh Size:........................

Group Categories of Fishes: **7 MISCELLANEOUS**

ISSCAAP Code	Division Group of Species	Local Name	English Name	Scientific Name	Weight (Kilogram)
76	**Sea-urchins,** Sea-cucumber and other echinoderms.	Landak laut, Trepang/ Balat	Sea-cucumbers (beche-de-mer)	*Holothuriodae*	
77	Miscellaneous Aquatic invertebrates	Ubur-ubur	Jellyfish	*Rhopilema spp.*	

Appendix 7: Fishes survey form: HA/kustem/cap5/5

Survey Form: HA/kustem/cap5/5

Date……………………………..

Location:………………………..

Fishing Gear…………………..

Mesh Size:……………………..

Division Group of Species	Local Name	English	Scientific Name	Weight (Kilogram)

HAMID AWONG FISHERIES MODEL (HAFM)

Appendix 8: Fishes survey form: HA/kustem/cap5/6

Survey Form: HA/kustem/cap5/6

Date……………………………..

Location:……………………..

Fishing Gear…………………

Mesh Size:……………………

TYPE OF FISH	WEIGHT (KG)
Commercial Fishes	
Trash Fish	
Others	
Total Catch	

Appendix 9 The length frequency of 24,653 samples

Standard length (cm)	Total Numbers
0	0
5	12
6	99
7	212
8	863
9	1123
10	1325
11	1530
12	1970
13	**2050**
14	1808
15	1675
16	1456
17	1693
18	**1795**
19	1521
20	1459
21	**1523**
22	1211
23	895
24	386
25	47
26	0

Appendix 10 The size distribution of *Priacanthus tayenus*

HAMID AWONG FISHERIES MODEL (HAFM)

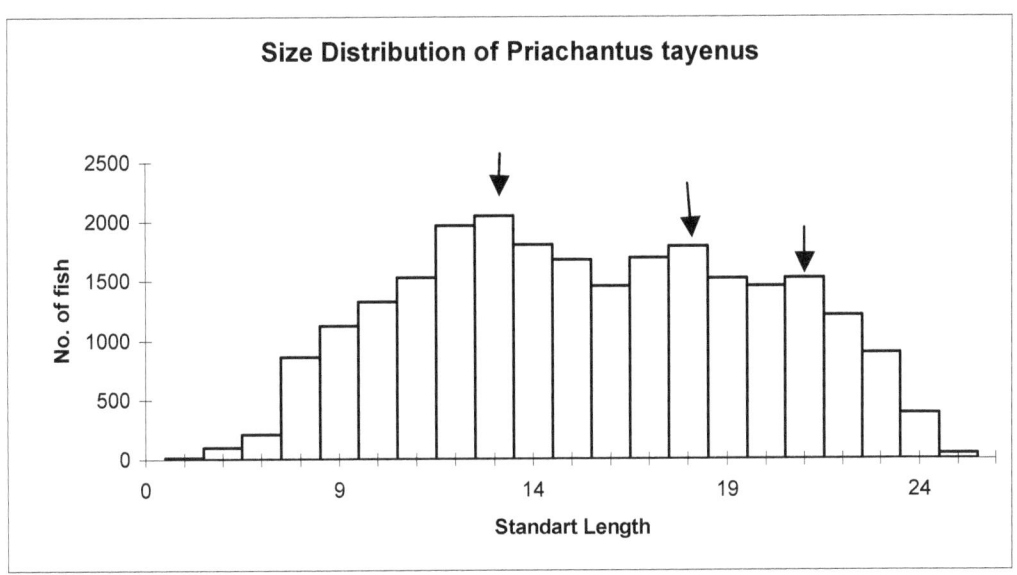

LISTS OF ABBERVATIONS

Short form

*	=	Multiplication
/	=	Division
a	=	Intercept
Anon	=	Anonymous
ANOVA	=	Analysis of Variance
B	=	Biomass
$B\infty$	=	Virgin biomass
b	=	Slope
C	=	Catch in numbers
$°C$	=	Celsius degree
CAY	=	Current Annual Yield
cm	=	Centimeter
CIA	=	Central Intelligence agency
Co-management	=	Cooperative management
CPUE	=	Catch per unit effort
DOF	=	Department of fisheries
E	=	Exploitation rate
Exp(x)	=	Exponential value of x
F	=	Fishing mortality
f	=	Fishing effort
FAO	=	Food and Agriculture Organization
Fig	=	Figure
g	=	Gram

GPS	=	Global Position Sensory
GRT	=	Gross tones
Ha	=	Hectare
L	=	Length
Lt	=	Liter
Ln(x)	=	Natural logarithm of x
ISSCAAP	=	International Standard Statistical Classification of Aquatic Animals and Plants
ITQ	=	Individual Transferable Quotas
K	=	Growth coefficient
kg	=	Kilogram
km	=	Kilometer
KO NELAYAN	=	Koperasi Nelayan Negeri Sabah
$L\infty$	=	Asymptotic length
LKIM	=	Lembaga Kemajuan Ikan Malaysia
Ln	=	Natural Log
Log	=	Logarithm
M	=	Natural Mortality
m	=	Meter
MAP	=	Marine Protected Area
MAY	=	Maximum Average Yield
MCY	=	Maximum Constant Yield
MEY	=	Maximum Economic Yield
mm	=	Millimeter
Ml^2	=	Miles Square
MSY	=	Maximum Sustainable Yield
Mt	=	Metric Tones

N	=	Number of Fish
NGO	=	Non Government Organization
pH	=	Potenz Hydrogen
ppt	=	Part per Thousand
ppm	=	Part per Million
PVC	=	Polyvinyl Chloride
q	=	Catch ability coefficient
R	=	Recruitment
RM	=	Ringgit Malaysia
RPM	=	Rotation Per Minute
S	=	South
s	=	Standard deviation
s^2	=	Variance
SAFMA	=	Sabah Fish Marketing
SD	=	Standard Deviation
SMPP	=	Sistem Maklumat Pengurusan Perikanan
t	=	Time at age
t_0	=	*Age at zero length*
TAC	=	Total Allowance Catch
N^o	=	North
v	=	Vulnerability
W	=	Weight
$W\infty$	=	Asymptotic weight
Y	=	Yield in Weight
Yr^{-1}	=	Per year
Z	=	Total Mortality

μm	=	Micrometer
%	=	Percent
φ	=	Phi
σ	=	Tau
q	=	Catchability Coefficient
$F_{0.1}$	=	Fishing Mortality at 10%
r	=	Correlation Coefficient

ABOUT OF THE AUTHOR

Hamid Awong was born on 9th December 1956 in Lahad Datu, North Borneo (Sabah, Malaysia). He attended his primary school in Native Volunteer School in Sapagaya, Lahad Datu on 1962, and then he was awarded a scholarship to attend secondary school in Melaka. Having completed secondary school he proceeds his matriculation course in Bogor Agriculture University, Bogor Indonesia and finally graduated Dip. II Seed Analyst.

The author has successfully completed his Post Graduate in Applied Science majoring in fisheries, in Australian Maritime College, Tasmania, Australia. This is the qualifying requirement for the Master in Applied Science (Living Marine Resources) in Australia and then successfully completed his Master of Applied Science majoring in Living Marine Resources. He has professional working experience as a researcher in Sabah Forest Development Authority (SAFODA), Kota Kinabalu, Sabah, Malaysia.

INDEX

A

age · 11, 12, 15, 20, 21, 22, 46, 113, 114, 115, 116, 117, 118, 119, 164, 165, 169, 178, 180, 181, 183
America · 4
Anatomy · 36
arenatus · 33, 34

B

Bara · 7
Brazil · 34

C

capita · 2, 16, 32
capture · 11, 12, 13, 21, 22, 35, 46, 111, 116, 117, 119, 165
Community · 30, 175
consumption · 1, 2, 7, 8, 15, 24, 32, 42
control · 1, 2, 5, 8, 12, 13, 21, 23, 27, 28, 43, 52, 169, 179
conversion · 30
cruentatus · 34

D

DARVEL BAY · 3, 7, 67
Distribution · 19, 34, 35, 37
dynamite · 3, 4, 13, 20, 44, 45, 169

E

East Coast · 5, 31, 67, 95, 177
economic · 2, 3, 4, 5, 6, 8, 9, 10, 11, 16, 20, 22, 23, 24, 26, 27, 28, 30, 31, 32, 43, 44, 45, 51, 52, 63, 70, 114, 168, 170, 175, 177, 178, 180, 184
economist · 3, 26, 31
Europe · 30

F

fish · 7, 1, 2, 3, 5, 6, 7, 8, 9, 10, 11, 12, 13, 14, 15, 16, 17, 18, 19, 20, 21, 22, 23, 24, 25, 27, 28, 29, 30, 32, 33, 34, 35, 37, 38, 40, 42, 43, 44, 45, 49, 50, 51, 52, 53, 56, 60, 61, 66, 71, 76, 84, 87, 88, 95, 96, 97, 99, 100, 102, 103, 104, 107, 108, 111, 112, 115, 116, 118, 123, 166, 167, 168, 169, 170, 172, 173, 175, 176, 177, 178, 179, 180, 181, 182, 183, 184, 191, 195
Fisheries · 4, 8, 1, 2, 3, 4, 5, 6, 9, 14, 20, 22, 27, 28, 29, 30, 31, 32, 41, 43, 44, 45, 46, 51, 52, 53, 54, 55, 56, 57, 58, 60, 71, 86, 90, 95, 96, 168, 175, 176, 177, 178, 179, 180, 181, 183, 184

H

Heteropriacanthus · 34

I

independent · 19, 30, 31
industry · 1, 12, 15, 21, 22, 31, 41, 45, 57, 64, 70, 96, 97, 170

K

kilograms · 7, 101, 104, 108

L

length · 11, 17, 28, 35, 37, 40, 82, 84, 86, 87, 91, 113, 114, 115, 118, 171, 172, 179, 182, 183, 184, 203, 204
Limiting · 28, 29, 30, 46, 47

M

MALAYSIA · 3, 7, 111
Management · 2, 3, 26, 27, 46, 51, 52, 121, 163, 175, 176, 177, 178, 180, 182, 184
manager · 2, 3, 4, 11, 12, 13, 44, 45, 96, 168, 169
marine · 3, 5, 9, 10, 12, 15, 16, 17, 19, 20, 24, 31, 41, 42, 43, 45, 46, 48, 49, 50, 51, 52, 53, 56, 57, 64, 65, 71, 96, 98, 104, 167, 168, 173, 175, 176, 178, 180, 181, 184, 195
Mexico · 33, 34
MSY · 1, 4, 26, 28, 44, 164, 168, 171

N

New Jersey · 34, 182
Nova Scotia · 34

O

own shore · 30

P

policies · 13, 21, 90, 165
Priacanthus tayenus · 7, 8, 9, 10, 11, 40, 64, 76, 82, 83, 84, 89, 90, 91, 111, 112, 113, 119, 121, 127, 136, 138, 146, 148, 157, 159, 161, 165, 173, 178, 204
projects · 30

R

redeployment · 30
reproductive · 1
Resources · 1, 9, 15, 99, 107, 175, 177, 179, 180, 181, 182, 184, 206
RICHARDSON · 3

S

Sabah · 4, 5, 7, 2, 4, 5, 6, 7, 8, 9, 14, 31, 44, 45, 46, 49, 54, 55, 56, 57, 58, 60, 63, 64, 65, 66, 67, 69, 70, 71, 73, 75, 86, 95, 96, 97, 98, 104, 108, 112, 122, 168, 170, 171, 173, 176, 177, 178, 183, 184, 206
schemes · 28, 30
Services · 4
species · 7, 1, 2, 3, 6, 8, 9, 10, 11, 14, 15, 16, 17, 18, 19, 29, 33, 34, 36, 37, 45, 47, 50, 51, 52, 60, 64, 67, 71, 76, 84, 87, 97, 98, 99, 101, 104, 108, 111, 113, 114, 115, 117, 118, 119, 122, 124, 146, 161, 163, 164, 165, 167, 168, 169, 171, 172, 173, 179
stock · 7, 8, 2, 3, 4, 6, 8, 9, 10, 14, 19, 21, 23, 26, 27, 29, 30, 31, 33, 44, 45, 47, 52, 64, 90, 96, 97, 99, 100, 107, 108, 111, 113, 116, 117, 122, 123, 124, 125, 127, 128, 129, 130, 131, 132, 133, 134, 135, 136, 146, 156, 159, 161, 163, 165, 166, 168, 169, 170, 171, 175, 176, 178, 179, 180, 182, 183, 184
sustainable · 1, 2, 3, 9, 10, 13, 27, 33, 43, 52, 90, 96, 111, 165, 171, 176
Sustainable · 1, 3, 4, 26, 28, 44, 45, 168, 171

W

weight · 11, 12, 17, 47, 85, 87, 97, 99, 100, 101, 105, 111, 115, 171

Y

years · 7, 8, 31, 42, 59, 64, 90, 91, 96, 108, 112, 113, 114, 115, 116, 117, 125, 127, 136, 146, 156, 159, 161, 164, 165, 166, 168, 171, 184

www.ingramcontent.com/pod-product-compliance
Ingram Content Group UK Ltd.
Pitfield, Milton Keynes, MK11 3LW, UK
UKHW051524180426
11947UKWH00018B/1569

9 780615 213217